新商业建筑
NEW SHOPPING CENTER

北京万创文化传媒有限公司 编　赵欣 译

大连理工大学出版社

图书在版编目(CIP)数据

新商业建筑:汉英对照 / 北京万创文化传媒有限公司编;赵欣译. —大连:大连理工大学出版社,2013.9
 ISBN 978-7-5611-8022-8

Ⅰ. ①新… Ⅱ. ①北… ②赵… Ⅲ. ①商业－服务建筑－建筑设计－世界－图集 Ⅳ. ① TU247-64

中国版本图书馆 CIP 数据核字 (2013) 第 149806 号

出版发行：大连理工大学出版社
　　　　（地址：大连市软件园路 80 号 邮编：116023）
印　　刷：上海锦良印刷厂
幅面尺寸：235mm × 300mm
印　　张：19
插　　页：4
出版时间：2013 年 9 月第 1 版
印刷时间：2013 年 9 月第 1 次印刷
策划编辑：袁　斌　刘　蓉
责任编辑：刘　蓉
责任校对：李　雪
封面设计：艾森品牌传播机构

ISBN 978-7-5611-8022-8
定　　价：280.00 元

电话：0411-84708842
传真：0411-84701466
邮购：0411-84703636
E-mail:designbooks_dutp@yahoo.com.cn
URL:http://www.dutp.cn

如有质量问题请联系出版中心：（0411）84709246　84709043

CONTENTS 目录

006
CRYSTALS AT CITYCENTER
水晶休闲购物中心

120
BEIJING SHIJINGSHAN WANDA PLAZA
北京石景山万达广场

018
SOLANA
北京蓝色港湾国际商区

138
EC MALL
北京欧美汇

056
SUZHOU TIMES SQUARE
苏州圆融时代广场

166
GINZA MALL
北京银座 mall

082
RAFFLES CITY
北京来福士中心

178
GLORY MALL
北京国瑞购物中心

108
MERCATOR CENTER MARIBOR
马里博尔墨卡托中心

188
CHINA CENTRAL PLACE
华贸中心

196
GEMDALE PLAZA
北京金地中心

246
YOU–TOWN LIFESTYLE CENTER
北京悠唐生活广场

208
VIVA
北京富力广场

256
JIN BAO PLACE
北京金宝汇

218
BEIJING APM
北京 APM

266
BEIJING YINTAI CENTER
北京银泰中心

226
INTIME LOTTE
北京乐天银泰百货

276
353 PLAZA
上海 353 广场

240
LONGDE PLAZA
北京龙德广场

294
CITYMALL
北京新中关购物中心

CRYSTALS AT CITYCENTER

水晶休闲购物中心

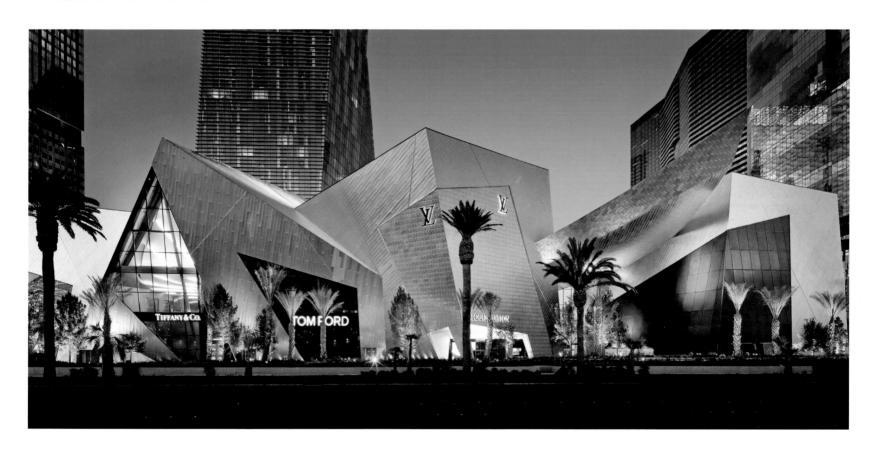

Location: Las Vegas, NV, USA
Rockwell Group's Area: 111,48.4 square meters
Gross Floor Area: 46,451.5 square meters

项目地址：美国内华达州拉斯加斯
罗克韦尔集团面积：11148.4 平方米
建筑总面积：46451.5 平方米

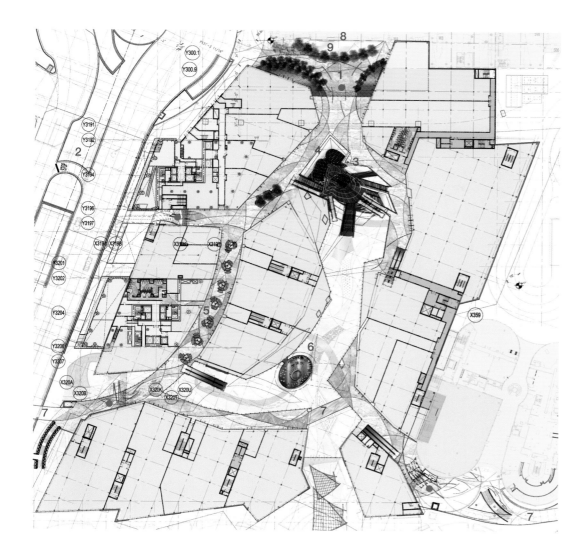

1 Casino Square
2 Casino Boulevard
3 Grand Stair
4 Pods
5 Hanging Gardens
6 Tree House
7 Park Bridges (3)
8 Casino Lobby
9 Reflective Pool
  Retail
  Entertainment, Food & Beverage
● Main Entrances (4)

Background

We designed the interior architecture for Crystals, the central retail, entertainment and dining district of CityCenter in Las Vegas, a joint venture between MGM MIRAGE and Infinity World Development Corp, a subsidiary of Dubai World. Crystals is the connective core of the hotels, resorts and residences that make up this new groundbreaking metropolis. Our firm joined a world-renowned group of architects who have contributed to this monumental urban project, including Daniel Libeskind, Cesar Pelli, Rafael Viñoly, Norman Foster and Helmut Jahn.

Design Concept

We envisioned Crystals as an abstracted 21st century park to fit into this new larger-than-life urban center in Las Vegas. The organic, curvilinear vocabulary of the interior architecture reinvents and re-imagines the idea of the central urban park as a social gathering place for shopping and dining.

Design Highlights

Interpretations and abstractions of nature animate the scenery that complements the sharp angles of Studio Daniel Libeskind's crystalline exterior, with a glowing three-storey wooden sculpture inspired by a modern tree house. Floor-to-ceiling columns and trellises above are abundant with hanging plants as a reinterpretation of a gazebo, and a dynamic flower carpet at the node of the three promenades in Crystals that transforms with the seasons. Natural and tactile materials abound in response to CityCenter's overall LEED Gold aspirations, including the flowing 7.3-meter grand bamboo stair, inspired by Rome's Spanish steps.

LEED Gold Details

Beyond the actual design, however, was the effort to use materials, techniques and processes that were in keeping with LEED Gold specifications.

• All the woods we used are FSC (Forest Stewardship Council) certified, including the sapele and mahogany used for the tree house, and the mahogany for the balcony rail and planters.

• We chose bamboo as the material for the grand stair that goes up 7.3 meters. Bamboo is a wood that easily replenishes, so we thought it would make sense to highlight it in a central design feature.

• Integrated walk-off grates – all entries will have integral walk off grates to help with indoor air, maintenance, and durability of finished surfaces.

• No added urea formaldehyde to wood products.

• All the substrates, sealants and the terrazzo flooring contain no toxic agents, and comply with all LEED criteria.

• Many of our structures use 100% recycled steel, including the hanging garden.

• Water efficient fixtures will achieve 38% savings over code, over 1.8 million gallons of water saved annually.

• All the integrated lighting is energy efficient LED systems.

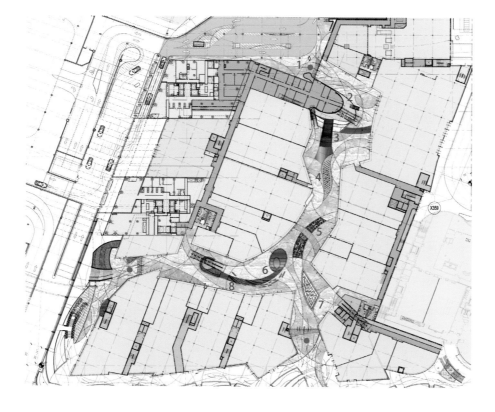

1 Crystals Valet
2 Oyster
3 Grand Stair
4 Water Vortex
5 Flower Carpet
6 Tree House Base - Concierge
7 Ice Totems
8 Tree Root
   Retail
   Entertainment, Food & Beverage
• Main Entrances (3)

背　景

建筑师设计了位于拉斯维加斯城中城的水晶休闲购物中心的内部空间，即中央零售、娱乐及餐饮区。水晶休闲购物中心是米高梅金殿梦幻和迪拜世界旗下子公司无限世界发展公司的合资企业。水晶休闲购物中心是连接这一新的开创性的大都市中的酒店、旅游胜地及居民区的核心。我公司与世界知名建筑师集团联手共建了这一卓著的都市项目，这些建筑师包括丹尼尔·里伯斯金、凯撒·佩里、拉斐尔·维诺利、诺曼·福斯特和赫尔穆特·扬。

设计理念

建筑师希望水晶休闲购物中心会成为一个抽象的21世纪公园，与拉斯维加斯这一具有传奇色彩的都市中心相映成趣。建筑内部有机、曲线的设计彻底改造并再次塑造了都市中央公园这一理念，使其成为购物及餐饮的社交场所。

设计重点

受现代树屋的启发，建筑师以一座熠熠生辉的三层木雕塑对自然进行了诠释和抽象描述，赋予了景致生命，弥补了丹尼尔·里伯斯金工作室设计的晶体外观的锋利棱角。从地板到天花板高的柱子和上面的格架攀爬了很多的悬垂植物，这是对露台的一种重新诠释，一张位于三个长廊交点处的充满活力的花毯则体现了水晶休闲购物中心的四季变化。为了达成购物中心的绿色环保建筑金奖的愿望，设计大量运用了自然和颇具触感的材料，包括流线型的7.3米长的大型竹楼梯，楼梯的设计灵感来源于罗马的西班牙台阶。

绿色环保建筑金奖细部

除去实际的设计，在材料选择、技巧运用及操作过程中，设计也遵循了绿色环保建筑金奖所要求的具体标准。

- 使用的所有木材都具有FSC（森林管理委员会）的证书，包括用于建造木屋的萨佩莱木和桃花心木，以及用于露台扶手及花架的桃花心木。
- 建筑师将竹子作为长达7.3米的大型楼梯的原材料。竹子易于深加工，建筑师认为在中央设计中强调竹子是明智之举。
- 协调一致的防护格栅——所有入口都有完整的防护格栅以保证内部的空气流通、维修保养及外立面完工后的结实耐用。
- 木制品中未添加尿素甲醛。
- 所有的基板、密封剂和水磨石地面均不含有毒物质，并遵循绿色环保建筑的标准。
- 包括空中花园在内，许多的建筑结构都是完全由回收的钢材制造而成的。
- 与用水标准相比，固定节水装置的运用将节省38%的用水，每年会节约180多万加仑的水。
- 所有的整体光源都采用节能的发光二极管系统。

New Shopping Center

SOLANA

北京蓝色港湾国际商区

Location: No.6 Chaoyang Park Road, Chaoyang District, Beijing
Total Site Area: 130,000 square meters
Gross Floor Area: 150,000 square meters
Architectural Design: Five Design Group, USA
Planning & Landscape Design: The Jerde Partnership
Landscape Design: Earthasia Design Group
Waterscape Design: FLUIDITY DESIGN CONSULTANTA
Lighting Design: Kaplan Gehring Mccaroll
Sign System and Environmental Design: Selbert Perkins Design
Developer: Beijing Blue Harbor Properties Co.,Ltd

项目地址：北京朝阳区朝阳公园路 6 号
占地面积：130000 平方米
总建筑面积：150000 平方米
建筑设计：美国 Five Design Group 公司
建筑规划及园林景观设计：美国 JERDE 国际建筑师事物所
园林景观设计：泛亚国际
水景设计：FLUIDITY DESIGN CONSULTANTA
灯光设计：Kaplan Gehring Mccaroll
标识系统及环境设计：Selbert Perkins Design
开发商：北京蓝色港湾置业有限公司

Solana is situated at the No. 6 Chaoyang Park Road, Chaoyang District, Beijing. It is composed of areas such as Merui shopping mall, SOLANA Mall, Liangma Gourmet Street, Dynamic City, Brand City and Liangma Wharf Bar Street. Its business covers retailing, catering, bars, boutique supermarket, cinema, skating club and electronic games places. The interior of the shopping park creates, with a completely open shopping space, a leisure and easy shopping atmosphere, and additionally, it integrates various business and advantaged environmental resources with the exquisite architecture. At the same time, it is also the first shopping mall that brings in the business mode of "Lifestyle Shopping Center".

On a land of one hundred and thirty thousand square meters, the developer only set up buildings with a gross area of 15,000 m^2, including nineteen European-style small buildings with two to three floors and business are carefully allocated to different areas. The landscape design is done by the Jerde Partnership and Earthasia Design Group. The designers have achieved a perfect match of shopping mall and the natural landscape of Chaoyang Park, breaking through the conventionality of designing landscape in shopping mall, instead, they designed a mall in a the landscape.

In such a water-lacking metropolis as Beijing, this project has the advantageous environment that it is surrounded by water from three sides. To the south and east, it is bordered by the vast lake of Chaoyang Park and to the north, it is the renovated Liangma river. The whole project seems to be a little town air-expressed from Europe, integrating commercial culture with natural landscape in a perfect way. The manual canal on the south bank extends the natural lake landscape, and the east-south lake bank has a large dropping water screen of two hundred and fifty meters in length, three meters in height, which creates a crystal palace-like visual effect. FLUIDITY DESIGN CONSULTANTA of the USA is in charge of the design of water landscape, and its design places emphasis on the relationship between the project and natural conditions, and each detail designed is planned according to the particular environment, meaning and value. There are not only the partial theme water landscapes created by moving points, but also integral water landscapes that flow through the whole project, so that it realizes the harmonious connection between sight and water landscapes and creates a special pro-water consumption experience for customers.

　　蓝色港湾国际商区位于北京朝阳区朝阳公园路6号。项目由美瑞百货、SOLANA MALL、亮马食街、活力城、品牌街以及亮码头酒吧街等区域组成，商业业态涵盖零售、餐饮、酒吧、精品超市、影城、滑冰俱乐部、电玩乐园等。商场内部以一种完全开放的购物空间来营造悠闲轻松的购物氛围，还将多样的业态、得天独厚的环境资源与精美的建筑相融合。同时，也是国内首个引入"Lifestyle Shopping Center"这一商业模式的商场。

　　蓝色港湾国际商区在130000平方米的土地上仅建设了150000平方米的建筑，共包括19幢洋溢着浓郁欧式风情的2至3层小型建筑，各种业态被精心地分配到各个区域。

　　该项目园林景观设计由美国JERDE国际建筑师事务所和泛亚国际联袂完成。设计师将商场与朝阳公园的天然景观完美地结合起来，打破了在商业中设计景观的常规思路，转而改为在景观中设计商业。

　　在北京这样一个缺水的大都市里，该项目却拥有三面亲水的优越环境，其南面和东面是朝阳公园的浩淼湖面，北面为经过整治的亮马河。整个项目仿佛是一个从欧洲空运而来的小镇，将商业文明与自然景观完美融合起来。南岸的人工运河延伸了自然湖水景观，东南湖畔拥有一个长250米、高3米的大型跌水水幕，营造出水晶宫般梦幻的视觉效果。美国FLUIDITY公司负责水景系统的设计，设计强调了项目与自然条件之间的关系，每个设计的细节都依据特定场所的环境、意义和价值而构思，既有在动线节点上创造的局部主题水景观，又有贯穿整个项目的整体水系，从而实现视觉与水景的和谐关联，为顾客营造了一份独特的亲水消费体验。

New Shopping Center

SUZHOU TIMES SQUARE

苏州圆融时代广场

Location: East of Jinji Lake, Suzhou Industrial Park (SIP)
Total Site Area: 210,000 square meters
Gross Floor Area: 510,000 square meters
Architectural Design: HOK International Co., Limited
Landscape Design: SWA Group

项目地址：苏州工业园区金鸡湖东
占地面积：210000 平方米
建筑面积：510000 平方米
建筑设计：美国 HOK 国际有限公司
景观设计：美国 SWA 景观设计事务所

Located at east of the Jinji Lake, Suzhou Industry Park, Suzhou Times Square is an "Urban Commercial Complex" and a one-stop complex commercial real estate project with large scale, modern realm and high quality, combining shopping, catering, recreation, business, culture and touring, etc. into one. In addition, it contains more than 4,000 underground parking spaces, six subway exits, a river shuttle service, an air corridor, and theme landscapes. Owing to these, Suzhou Times Square has become the most prosperous business area within the CBD circle in Suzhou and the most influential and valuable commercial street in East China region.

Through many a air corridor, the whole Suzhou Times Square is connected, so that consumers can enjoy both the interior and exterior leisure views from different perspectives. According to the shopping purposes of consumers, Suzhou Times Square is divided into five functional areas to fulfill shopping needs of different consumers, and at the same time, it serves as an effective distribution to people inside of the shopping square.

Tianmu Street: Composed of many a individual building in an organic way, here, consumers can enjoy more than one hundred brand retailing outlets. The world first five-hundred-meter-tall LED screen is consisted of more than twenty million UHB LED lights that cost a hundred million RMB.

Living and Leisure District: Aimed to promote delicate life style and fully satisfy family consumptions, the district collects theme malls related to home furnishings, home appliance, digital and Children's world etc.

Jiuguang Department Store District: covering an area of 170,000 square meters, this district injects metropolis charms to Suzhoua. By far, it is the largest commercial department flagship store in Suzhou as well as in China. With international and domestic top brands and their fashionable products, targeting at the middle and high-end classes, this district provides more chances for Suzhou consumers to be in line with the international fashion trends.

Waterfront Food District: Global cuisine and enjoyment are collected in this district.

Business Office District: Four international standard office buildings form a international business complex which is also the landmark of CBD area.

圆融时代广场位于苏州工业园区金鸡湖东岸，是集购物、餐饮、休闲、娱乐、商务、文化、旅游等诸多功能于一体的大规模、现代化、高品质的"城市级商业综合体"及一站式消费的复合性商业地产项目。项目包括4000多个地下车位、6个地铁出口、水上巴士、空中连廊、主题景观等，成为了苏州市区CBD最繁华的商业中心和华东地区最具影响力和商业价值的品牌街区。

整个时代广场由多个空中连廊连通各个区域，消费者可以从不同角度观赏室内外的休闲景观。该项目内部按消费者的购物目的分为五大功能区，满足了消费者不同的购物需求，也有效地分流了购物广场内部的人群。

圆融天幕街区：由多个独立建筑有机串联而成，在这里，消费者将能够接触到100余家品牌零售店。其世界第一的500米巨型神奇天幕由2000多万个超高亮度的LED灯组成，耗资亿元。

生活休闲区：包括家居、家电、数码及儿童天地等主题专业商场，倡导精致生活方式，满足家庭式消费。

久光百货区：17万平方米的苏州久光百货为苏州注入了时尚都会魅力，是目前苏州、也是国内单一百货店最大面积的旗舰商业航母，汇聚了世界及国内一流的品牌时尚潮流产品，定位中高端，让苏州消费者与更多国际时尚潮流相接轨。

滨河餐饮区：汇集环球无国界美食。

商务办公区：四栋国际标准写字楼组成了CBD地标级的国际商务建筑群。

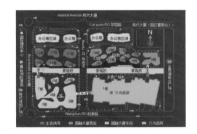
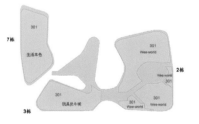
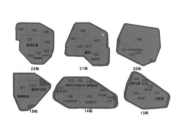
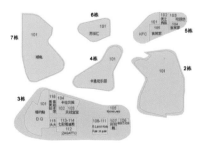

RAFFLES CITY

北京来福士中心

Location: Southwest Dongzhimen Overpass, Dongcheng District, Beijing
Total Site Area: 14,685 square meters
Gross Floor Area: 145,928 square meters
Shopping Space: 29,716 square meters
Architectural Design: Stephen Pimbley
Developer: Capitaland

项目面积：北京市东城区东直门立交桥西南角
占地面积：14685 平方米
建筑面积：145928 平面米
商业面积：29716 平方米
建筑设计：Stephen Pimbley
开发商：凯德置地

Raffles city is the first large-scale multi-functional building of Capitaland in Beijing, which is also the third world-wide architecture of Raffles brand in the series by this group. This project is located in the land of Dongzhimen, which is the core area of business district in East Second Ring, Beijing. It faces Dongzhimen Nei Avenue on the north, adjoins East Second Ring Road on the east, and borders the second embassy area, with numerous office buildings, large-scale malls, hotels and cultural centers. This project connects with the largest modernized integrated transport hub and Airport Express Line in Asia. Located in the great rendezvous of corporations' headquarters, it is adjacent to CNOOC, Sinopec, China Telecom and the rest 11 business elites in China, with a total of about 150,000 square meters gross floor area, it is composed of a first-rate office building that occupies a floor space of 41,000 square meters, 265 sets of apartments with top-level service and a deluxe shopping mall covering an area of 30,000 square meters as well as over 610 parking spaces.

Raffles city has been completed by the end of 2008, of which the commercial portion is put into trial operation in April, 2009. This project saw the introduction of multitudinous top brands, such as Watson's, ZARA, Loccitane, WHM Chinese Meal, M)Phosis, Frank Provost and Spectacle Hut.

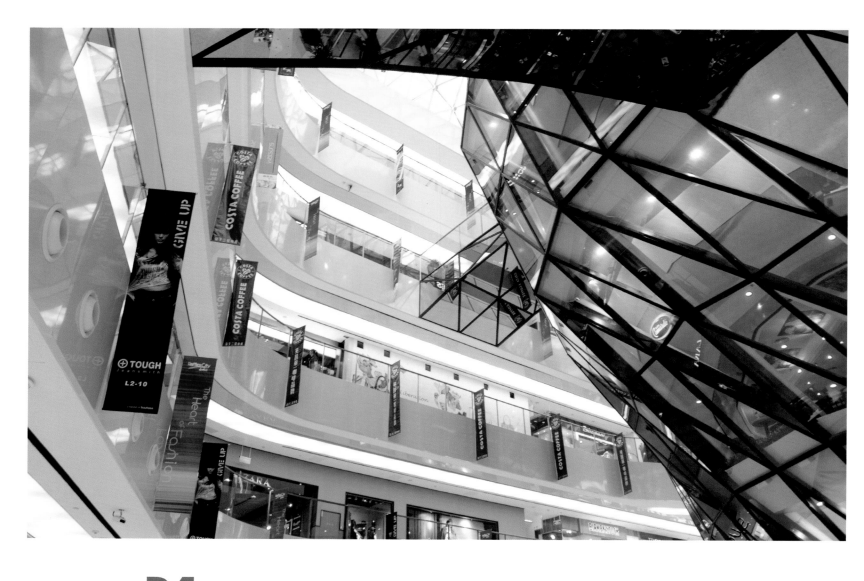

B1 荟珍阙
Gourmet Hall

L1 明星汇
Celebrity Walk

来福士中心是凯德置地在北京的第一个大型综合建筑,也是该集团在全球范围内的第三座"来福士"品牌系列建筑。该项目位于北京东直门国门之地,东二环的核心商圈地带,北临东直门内大街,东接东二环路,紧邻第二使馆区,周围遍布写字楼、大型购物中心、酒店和文化中心。它连接亚洲最大的现代化综合交通枢纽和机场快线轨道,身处企业总部云集之地,并与中海油、中石化、中国电信等11家中国经济巨头为邻。该项目总建筑面积约150000平方米,由一栋41000平方米的甲级写字楼、265套顶级服务公寓、约30000平方米的豪华购物中心以及超过610个停车位组成。

来福士中心2008年底竣工,商业部分于2009年4月投入试运营。项目引进了Watson's、ZARA、Loccitane、WHM Chinese Meal、M)Phosis、Frank Provost、Spectacle Hut等众多顶尖品牌。

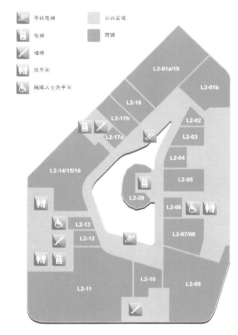

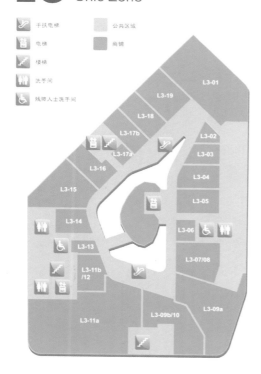

L3 潮尚地
Chic Zone

L4 动感界
Chill Out Place

New Shopping Center | 新商业建筑

L5 百味坊
Chef's Choice

MERCATOR CENTER MARIBOR

马里博尔墨卡托中心

Location: Maribor, Slovenia
Building Area: 41,000 square meters
Architect: Andrej Kalamar
Design Company: Studio Kalamar
Client: Mercator d.d.

项目地址：斯洛文尼亚马里博尔
建筑面积：41000 平方米
建筑师：Andrej Kalamar
建筑公司：Studio Kalamar
客户：Mercator d.d.

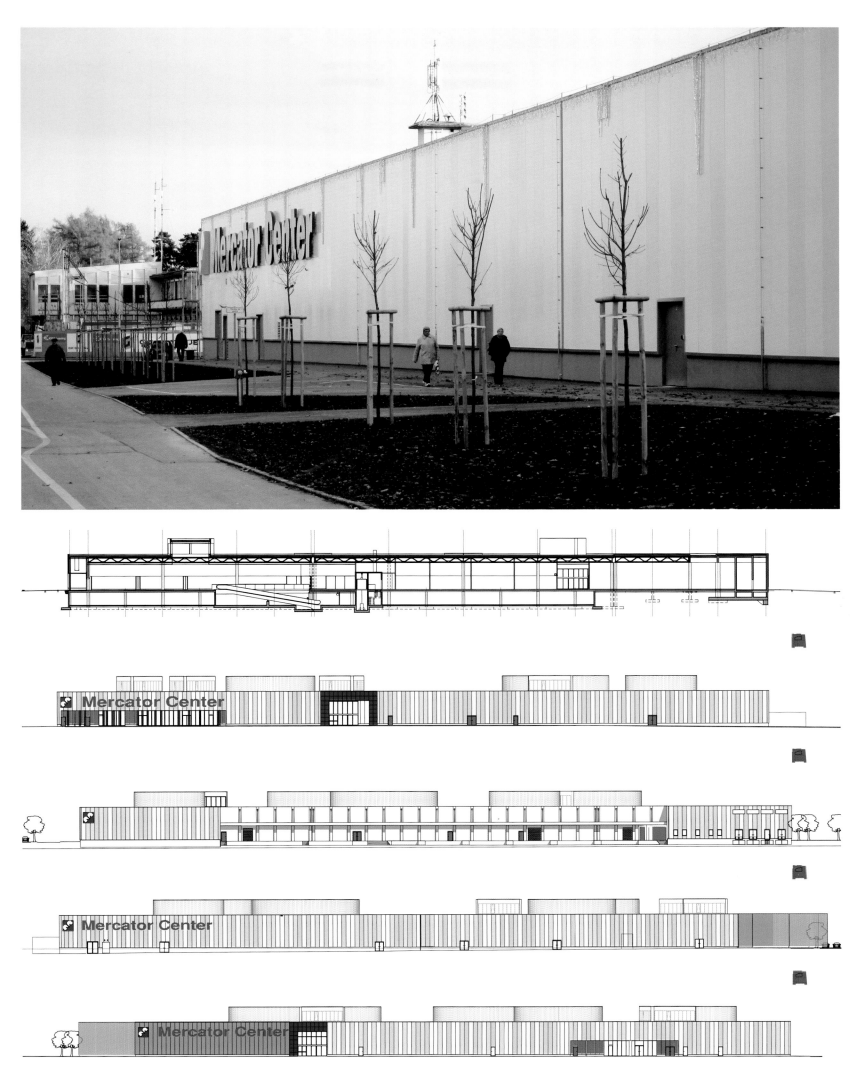

Mercator Center Maribor is located in a densely populated part of Maribor, which distinguishes it from typical suburban retail centers. The large center is therefore more open, has several entrances, oriented not only to visitors who come by car but also to pedestrians. Entrances are typologically accentuated. They lead into interior malls and squares, creating an urban experience of a real city.

With the uncontrolled versatility of visual impulses from the shops left and right we devoted our attention to design of the floor and ceiling, which bring some architectural articulation into the diversified space. The floor is defined by sand-colored ceramics with a diagonal darker path leading from one side to the other and back. This space is bordered in black, advising that beyond the border we cross into heterogeneous impulses of shops. Even more expression is introduced in the ceiling, which is transformed from a surface into volume with the use of lighting fixtures, developed especially for the project. Ceiling reflects the dynamic of motion through the forest. Each light is the size of an adult man; the dynamics of drafted paths leads them through the entire building. Thus, the mall is always full, giving a good sensation to the visitors. This expression was reached in a very empirical way: the inspiration came from the question of how to reduce the use of energy needed to operate the building. The level of light was lowered for 1.80 meters, the size of tubular lights. This reduced the energy necessary for illumination by 20% compared to lights fixed directly to the ceiling.

As the center is surrounded by high rise buildings, its fifth facade – roof also plays a significant role. The entire roof is green, providing adjoining residents with the view of a park as well as preventing the effect of heat island. The large amount of HVAC equipment on the roof would normally provide a visual disturbance as well a source of noise; these disturbances are neutralized by enveloping the equipment with large wooden cylinders. Positioned on the green roof, the cylinders function as pavilions freely distributed in the park.

Contrary to general practice in our cultural environment, the client briefed us in advance of the limited construction budget and expectations of low energy consumption. This has turned out to be a challenge of constant search for creative solutions – we prepared projects for building that set new standards regarding investment value and running costs. Sustainable design was not understood as a restriction, but as a source of inspiration.

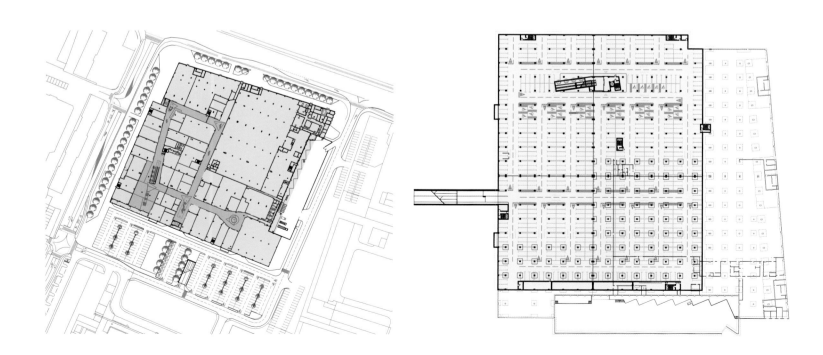

New Shopping Center | 112-113 新商业建筑

　　马里博尔墨卡托中心位于马里博尔人口稠密的地区。不同于典型的郊区零售中心，墨卡托中心更为宽敞，设有多个入口，不仅适用于驾车的顾客进入，还适合步行之人。入口按类别划分，通向内部购物商场和广场，营造了一种真正城市意义上的都市体验。

　　左右两侧的商店带给我们无限的视觉冲击，令建筑师们注意到了地面和天花板的设计，它们为这个多元化的空间带来了更多的建筑元素。地面选用的是沙滩色的地砖，一条颜色更深的曲折的小径将顾客引入另一侧和后面的空间。空间镶以黑色边，象征着越过边线，顾客便进入了林林种种的店铺之中。天花板被赋予了更多的表情，利用特别为此项目设计的照明灯具使其从平面变成了立体造型。天花板反映了在森林中穿行的流线。每盏灯都有成人般大小；富有动感的多条小径贯穿了整个建筑。因此，购物商场总是人流不息，令人叹为观止。设计结合了实证研究：设计的灵感来自于对如何降低运营该建筑所需的能源的考虑。灯的高度降到了1.80米，并采用了管状灯。与将照明灯直接固定在天花板上相比，这种做法节省了20%的照明所需能源。

　　由于此购物休闲中心四周均有高层建筑，因此，其第五立面，即顶部，起到了至关重要的作用。整个顶部绿色环保，使附近的居民既能看到公园的景致，又避免了热岛效应。通常在屋顶上使用大量高压交流电设备不仅会造成视觉干扰，还会形成噪音源；设计通过在这些设备外套大型的木制圆柱，中和掉了这些干扰。绿色环保屋顶上的这些圆柱自然分布在公园中，还起到了凉亭的作用。

　　与一般文化环境下的常见做法不同的是，客户提前告知建筑师他们的建筑预算有限，希望做到低能耗。因此，这给建筑师提出了一个挑战，需要他们不断地去寻求创造性的解决方法——建筑师准备了多个建筑方案，为投资价值和运营成本建立了新标准。可持续设计在此并不是一种限制，反倒成了灵感的来源。

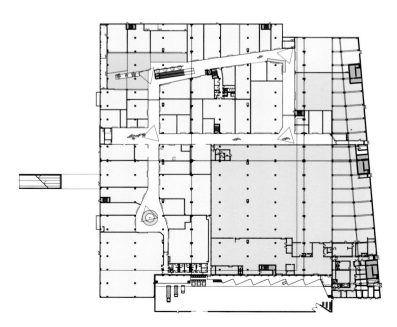
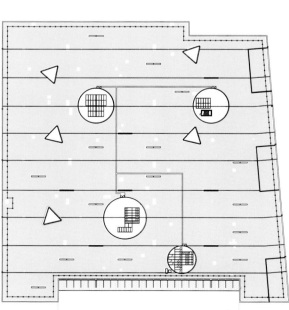

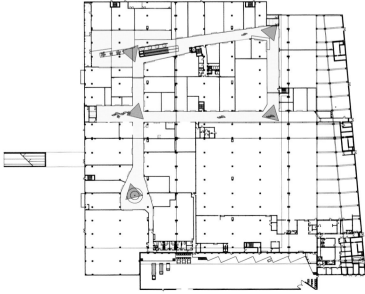

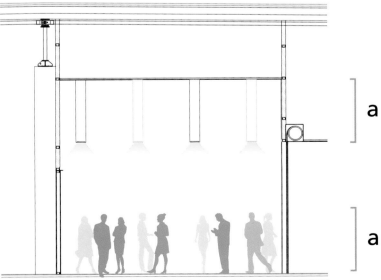

New Shopping Center

BEIJING SHIJINGSHAN WANDA PLAZA

北京石景山万达广场

Location: Northwest Corner of the Intersection of Shijingshan Road and Lugu Street, Shijingshan District
Total Site Area: 70,000 square meters
Gross Floor Area: 280,000 square meters
Architectural Design: M.A.O Masters Architectural Office, Beijing Institute of Architectural Design
Developer: Dalian Wanda Group Co., Ltd.

项目地址：石景山区石景山路与鲁谷大街交汇处西南角
占地面积：70000 平方米
总建筑面积：280000 平方米
建筑设计：日本 M.A.O 一级建筑士事务所、北京市建筑研究院
开发商：大连万达集团

Beijing Shijingshan Wanda Plaza is located in the extension cord of Chang'an Street, occupying the core area of Shijingshan District, which is so far the only large-scale mixed-use project near Shijingshan District and West Chang'an Street. The project covers an area of seventy thousand square meters, with a total gross floor area of two hundred and eighty thousand square meters, planned with various urban functions, such as top-grade hotel, 5A class office building, international supermarket chains, multi-functional theater, large-scale entertainment center, high-end department store, pedestrian mall and comprehensive dining and recreational center.

The overall planning of Beijing Shijingshan Wanda Plaza integrates many factors perfectly, such as construction culture, environmental impact, and the application of materials, emphasizing the relationships between elements, for instance, "human" and "building", "facade" and "image", so that the intersection between them will produce a new building structure language.

The project makes use of similar graphic patterns and formation mode to show the qualities of various kinds of architectural forms, so that the facades of the building can not only express a unified character, but also demonstrate the temperament of each building. The project features a solemn and calm, noble and elegant connotation, highlighting the characters of the area where it is located – Beijing + capital, Chang'an Street + Shijingshan Road – continuing its history, stressing on its cultural accumulation and the traits of realistic street space.

In this project, large glass surfaces are used as warp threads, vertical lines are used as weft threads, the warp and weft threads are intertwined. The body of the building is designed with concatenation of unified and flowing facades, which makes the group image more integrated.

The application of high-rise monomer interludes and groups with the same elements form the overall image of the building, coupled with the combination of continuous moving lines on the lower business part of the building, creating an integrated and generous feeling for the whole project no matter how one's view and perspective changes.

The reversal and interludes of the building body, the changes of surface texture, and the conversion of various spaces and times create a fresh "Wanda new space."

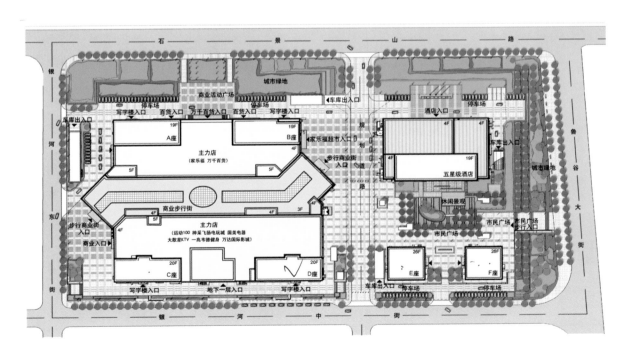

　　北京石景山万达广场坐落于长安街延长线上，占据着石景山的核心区域，是目前石景山区、西长安街附近唯一的大型综合体项目。该项目占地 70000 平方米，总建筑面积 280000 平方米，规划有高档酒店、5A 级写字楼、国际连锁超市、多功能影院、大型娱乐城、高档百货商场、步行商业街和综合餐饮娱乐中心等多重城市功能。

　　北京石景山万达广场整体规划将建筑文化、环境影响、材料运用等多种要素进行了完美统一，重点突出了"人"+"建筑"、"立面"+"印象"的要素关系，使两者的交集产生了新的建筑构成语言。

　　该项目运用相似的图形模式、构成模式来表现各种不同建筑形态的气质，使建筑外观整体统一，同时还彰显了各建筑不同的气质。项目庄重平和、高贵典雅的气质内涵，突出了所处区域——北京+首都、长安街+石景山路的历史延续、文化积淀和现实的街道空间特点。

　　该项目以大块的玻璃表面作为经线，以竖向线条为纬线，经纬相互交织。建筑的体块由统一流动的立面气质串联起来，使其群体形象更加统一。

　　高层单体的体块穿插和相同要素的群体应用所形成的建筑整体形象，配合低层商业部分连续动线的组合，使建筑在不同的观察角度、不同的视点上都会产生统一大气的感觉。

　　建筑体块的翻转、穿插和表皮纹理的变化、多种时空的转换，创造出全新的"万达新时空"。

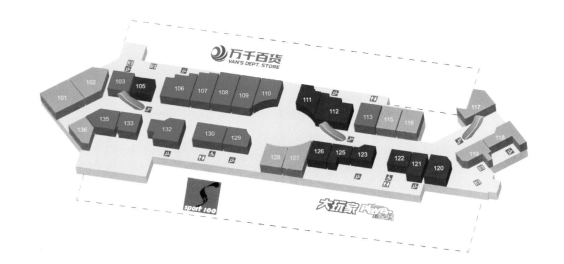

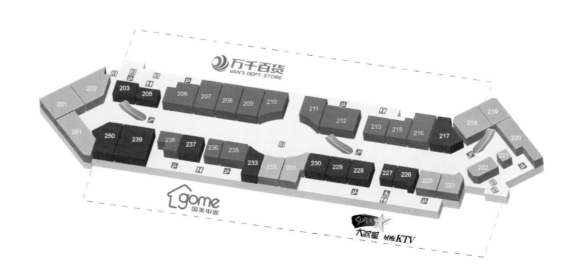

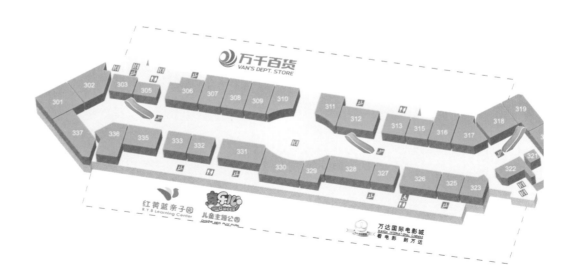

EC MALL

北京欧美汇

Location: Zhongguancun West District
Architectural Design: Benoy
Developer: ECM

项目地址：中关村西区
建筑设计：Benoy
开发商：ECM

The historic Haidian Xie Street(the pedestrian street on Zhongguancun Square), the rich heritage of the ancient culture reserved in the strong modern commercial atmosphere, what EC Mall brings to consumers is a unique feel and experience, with convenient traffic as the added brilliance to the splendor of this project. The many internationally-known trend fashion brands entering into EC Mall also enhances infinite charm to it. Integrating shopping, catering, recreation, and amusement into one, EC Mall pioneers the shopping enthusiasm of Beijing, especially Zhongguancun.

The general introduction to EC Mall: six floors above the ground, one floor under the ground, 52,471 square meters of business retail space, and 529 parking spaces in the underground parking lot. In addition, dynamic large screen and free WIFI are also available.

EC Mall is located in the NO. 23 Land of the west district in Zhongguancun. This project is designed by Benoy, an internationally-known architectural design company, whose design philosophy-"the symbiosis of creativity and vitality" has a far-reaching influence on the design of today's commercial architecture.

In the interior design of EC Mall, based on adhering to the traditional concept, Benoy places its emphasis on the close connection between architecture and nature, which further explores the meaning of life extension; combining high technology with modern sense, three large-scaled open hollow interior spaces bring a feel of vast and transparent. Along with the spacious skylights and the glass courtyards on two sides of the shopping square, EC Mall appeals to people with its remarkable design, and make them cannot bear to part this distinctive shopping feel. The outer wall of the architecture and the large decorative lighting in the interior space bring out the best in each other, which upgrades the class of the whole shopping mall, and complements the cityscape of the whole Beijing.

历史的海淀斜街（中关村广场步行街）、古老的文化积淀蕴藏在浓厚的现代商业气息中，带给消费者的是独特的感触与体验，便利的交通条件更为欧美汇购物中心锦上添花。众多国际知名潮流时尚品牌的入驻也为欧美汇购物中心增添了无限魅力。集购物、餐饮、休闲、娱乐为一体的欧美汇引领了北京，尤其是中关村地区的购物热潮。

欧美汇概况：地上 6 层，地下 1 层，共 52471 平方米的商业零售空间，地下停车场拥有 529 个停车位，此外，还有动感大屏、全场免费无线上网。

欧美汇位于中关村西区 23 号地块，该项目的建筑设计由世界著名建筑专业设计公司 Benoy 负责，其"创造力与生命力共生"的设计哲学对当今商业建筑设计产生了深远影响。

在欧美汇的室内设计上，Benoy 在秉承其传统理念的基础上，强调了建筑与大自然的密切联系，进一步探索了生命延伸的意义；将高科技与购物中心的现代感融和在一起，三个大规模开放式的中空空间给人以辽阔通透之感，再配上开阔的天窗及购物广场两侧的玻璃中庭，使顾客为这超凡的设计所吸引，留恋于此般独有的购物感觉。建筑外墙与室内的大型灯饰相得益彰，提升了整个商场的格调，与整个北京的城市风情相互辉映。

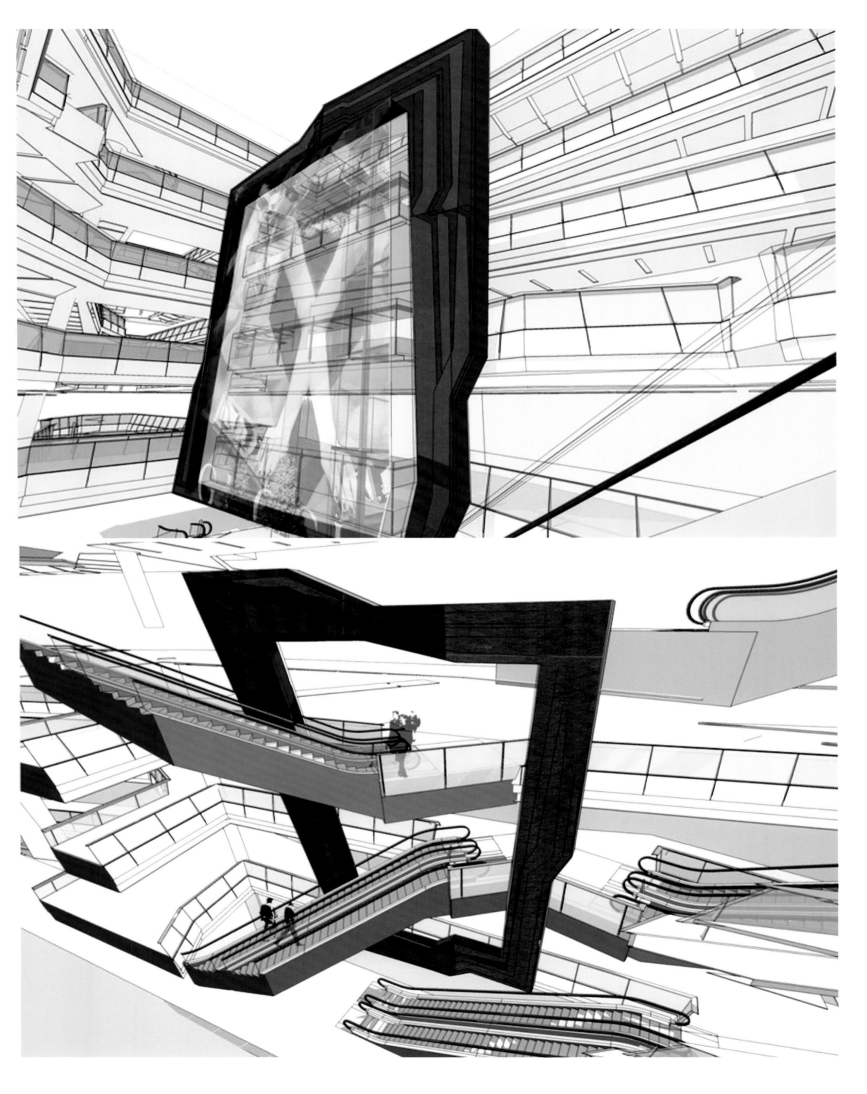

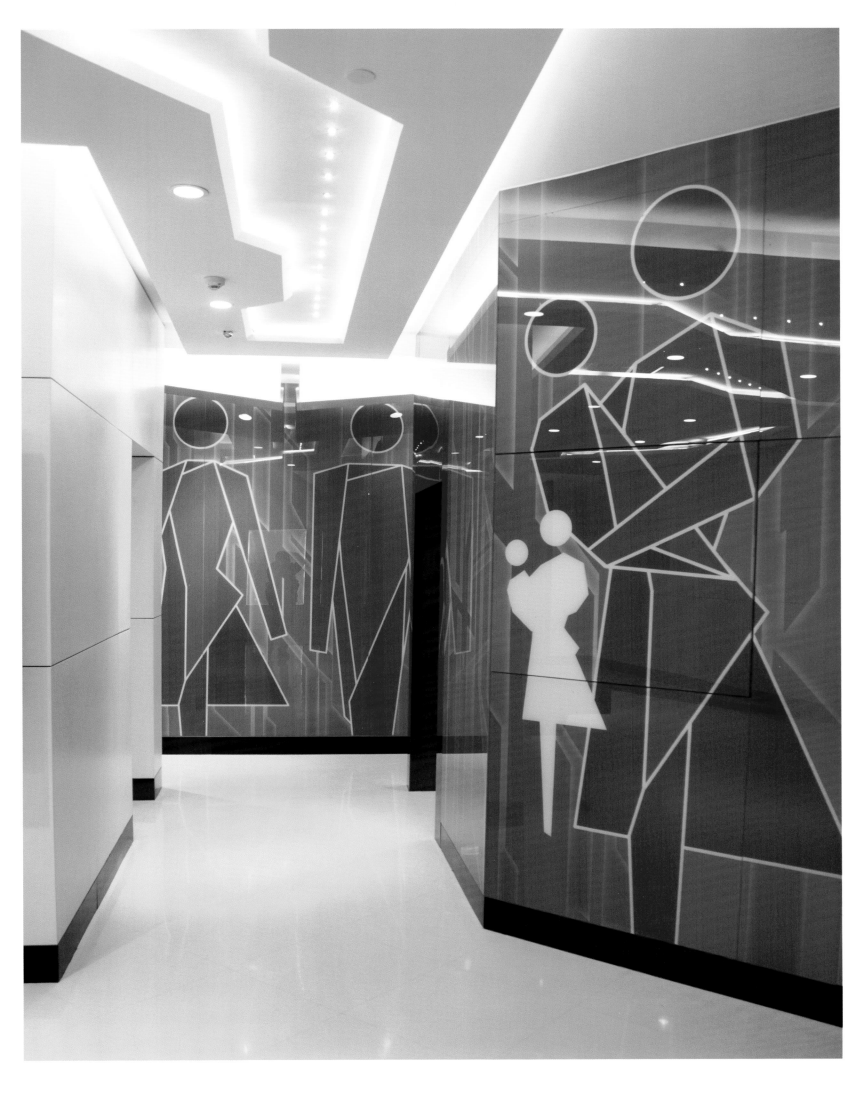

GINZA MALL

北京银座 mall

Location: No.48 Dongzhimen Wai Street, Dongcheng District, Beijing
Developer: MTR Corporation Ltd

项目地址：北京市东城区东直门外大街 48 号
开发商：香港铁路有限公司

Depending on the rich shopping mall operation experience of MTR Corporation, Ginza Mall injects new fashion elements to the capital city Beijing. The capacious shopping space and the design that highlights the shopping image have successfully attracted more than fifty well-known domestic and overseas clothes and shoes brands, over twenty food-and-beverage shops, thirty skin care and cosmetics brands and accessories brands, it provides a one-stop shopping experience place for young people especially young ladies, in Beijing who walk at the forefront of the times. Ginza Mall is situated at the Exit C of Dongzhimen Metro Station, providing a thoughtfully convenient shopping paradise for Beijing residents who lead a busy life.

The overall positioning of Ginza Mall is high-end and mid-end groups, targeting mainly at the mid-level white-collars, on the basis of which each floor of the mall set types of business and brands appealing to customers of different levels. For instance, fifteen gourmet brands are distributed in each floor: B1 and B2 floors give their priority to simple dining restaurants, and are white-collars-oriented; the second and the third floors have CHITOSE, PacificCafe and so on, and are management-oriented.

Ginza Mall covers many types of business, including garment, accessories, shoes and bags, cosmetics, clocks and watches, household appliances, and home furnishings. It has many brands, for example, i.t., Levi's, Fornarina, Ole, I.S.O.. In addition, for brands like Bunka, I.S.O. and Very, it is the first time for them to enter the Chinese market.

Basement 1
地下一层

Basement 2
地下二层

　　银座mall凭借港铁公司丰富的商场运营经验,为首都北京注入了新的时尚元素。宽敞的购物空间以及突显商铺形象的设计成功地吸引了超过50个的国内外知名服装、鞋履品牌,逾20家餐饮店铺,以及30个护肤品、化妆品及饰品品牌,为北京走在时代尖端的年轻人,特别是女士提供了一个一站式购物体验之所。银座mall位于东直门地铁C出口,为生活繁忙的北京市民提供了贴身的便利购物天堂。

　　银座mall整体定位为中高端,主要面向中层白领阶层,在此基础上各楼层会针对不同层次的消费者设置业态和品牌。比如,15个餐饮品牌分布在各个楼层,其中B1和B2层主要以简便的餐饮食肆为主,主要面向白领员工;2层和3层有千登世日本料理、PacificCafe等,主要面向管理层。

　　银座mall涵盖了服装、饰品、鞋包、化妆品、钟表、家电、家居等多种业态,包括i.t.、Levi's、Fornarina、Ole、I.S.O.等多个品牌,其中Bunka、I.S.O.、Very等更是首次进驻北京。

Level 3
三层

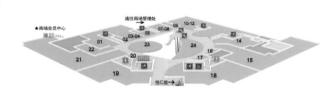

Level 2
二层

Level 1
一层

New Shopping Center

GLORY MALL

北京国瑞购物中心

Location: No.18 Chongwenmen Wai Street, Beijing
Landscape Design: Belt Collins International (HK) Limited
Interior Design: Concept I Design Co., Ltd.
Developer: Glory Real Estate

项目地址：北京崇文门外大街 18 号
景观设计：贝尔高林国际（香港）有限公司
室内设计：国际概念设计有限公司
开发商：国瑞地产

Glory Mall is by far the largest commercial project in Beijing owned by Glory Real Estate. It is located in the Chongwai Business Circle inside of the Second Ring, which is by the side of Chang'an Street. Targeting as an elegant and vigorous one-stop theme shopping center, this project integrates hotel, international finely decorated apartments, 5A top class office building into one, covering an construction area of up to one hundred and thirty thousand square meters.

At the very beginning of the planning, it took full consideration of the home shopping environmental quality and the green and cozy sensational enjoyment. Therefore, the shopping center is fresh and bright in its decoration, and the shopping circulations are rich and solid. Curve elements are most frequently used in the project, for instance, the question-mark shaped LED consulting desk, the curve LED suspending ceiling, the multi-functional fountain stage floating on water as well as the unique large five-storey high waterfall, water screen sight-seeing elevators. The interior of the building designs green scenic sight as much as possible. Not only in terms of sight, but also in terms of hearing, one can both find the rare flowing sound of water in such a steel-concrete city. The interior of the building adopts vigorous lights to set off characteristic things, and through the application of fun geometrical shapes and symbolic signs, it creates recognizable symbolic points, perfectly combining the practicality and aesthetic value.

Glory Mall has two sight-seeing lifts, thirty six elevators, four cross-floor elevators, six goods elevators and five passenger elevators. By improving horizontal and vertical transportations, it upgrades the circulations to all floors, making them simple, smooth and effective. The Triumphal arch typed bridge in the atrium can gather the flow of people in the commercial plaza in an effective way. In addition, the design strives to maximize the size of each shop through carefully shaping the heightened space's form and size.

Furthermore, Glory Mall has done a deep and scientific research on the combinations of brands and proportions of different industries.

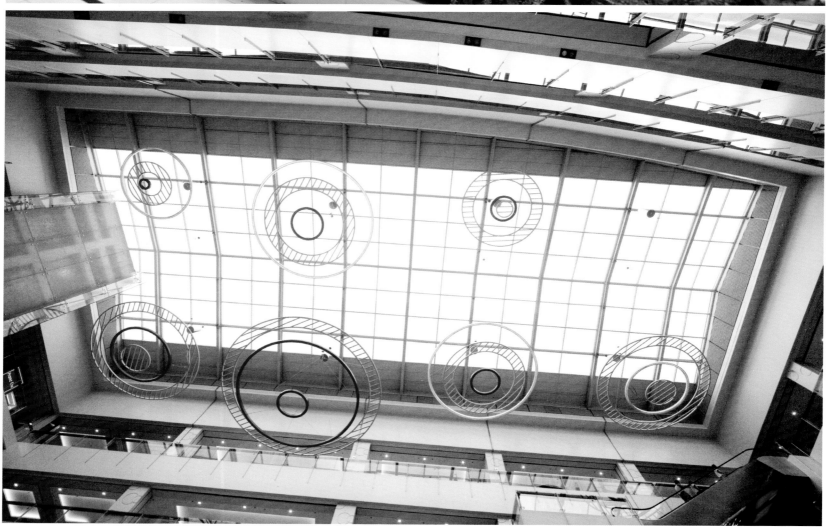

　　国瑞购物中心是国瑞地产旗下目前在北京最大的商业项目。项目地处长安街畔、二环内的崇外商圈，定位为典雅而充满活力的一站式主题购物中心，融酒店、国际精装公寓、5A甲级写字楼于一体，总建筑面积达130000平方米。

　　建筑在设计之初就充分考虑到了家庭购物的环境品质与绿色温馨的感官享受，购物中心的装修风格清新明快，购物动线丰富立体。曲线元素是设计最常用的元素，设有问号形的全LED咨询台、曲线形LED吊顶设计、多功能水上喷泉舞台以及北京独一无二的落地五层的大型流水瀑布、水幕观光电梯等。建筑内部尽可能多地设计了绿色景观，不仅是视觉，在听觉上，也会让人们找到那种在钢筋混凝土的都市中少有的潺潺水声。建筑内部运用活力四射的灯光来烘托特征物，通过有趣的几何图形及标志符号的运用，营造易于识别的标志节点，将实用性和观赏性完美结合。

　　国瑞购物中心共拥有景观电梯2部、扶梯36部、跨层扶梯4部、货梯6部、客梯5部，通过提升水平和垂直交通来优化通往所有楼层的动线，使之简洁、流畅、高效。中庭的凯旋门式过桥，能够有效地汇聚商场人流。另外，通过精心地塑造挑空空间的形状和尺寸，设计还尽可能地使每一家店铺的面积都达到最大化。

　　此外，国瑞购物中心在品牌的组合规划及各业态间的配比上都进行过了深入、科学的研究。

CHINA CENTRAL PLACE

华贸中心

Architectural Design: Kohn Pedersen Fox Associates (KPF), UAS , East China Architectural Design & Research Institute Co.,Ltd
Landscape Design: EDAW, USA
Hospitality Design: WATG, USA
Developer: Guohua Property Management Co.,Ltd

建筑设计：美国 KPF、华东建筑设计研究院
园林景观设计：美国 EDAW
酒店设计：美国 WATG
开发商：北京国华置业有限公司

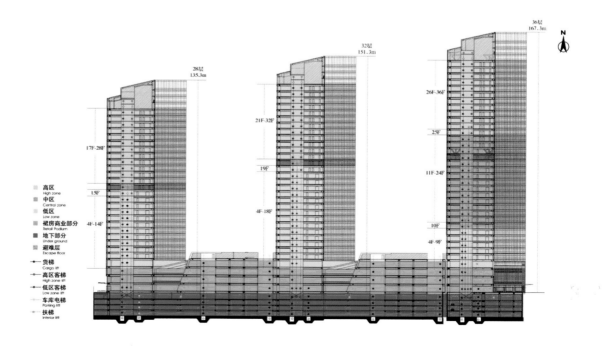

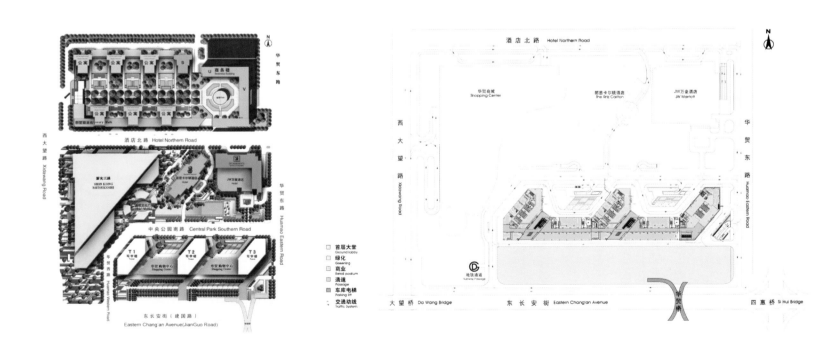

China Central Place is one of the sixty major projects in Beijing City in 2003; so far it is the only building complex covering an area of millions of square meters and goes across the East Chang'an Street. It faces Beijing CBD core area on the west, integrated into and became a part of the CBD. Coming into Chang'an Street from Guobin road on Airport Road, you will be first greeted by China Central Place.

The overall design of the project is made by the world-renowned KPF Architects, and the architectural design is pure, simple, bright, and sharp with strong sense of memory and times. Based on the concept of "combination of architecture and urban culture, interdependence of architecture and urban environment", the concept of the urban design in the overall plan is to link the west and east sides of the plot, so that the business zone naturally becomes sparse from south to north. The designing purpose of KPF Architects is to create a series of urban space which is composed of building complex, and to form a necessary link between the appearance and existence of the urban space and its surrounding environment.

The project features three super-5A intelligent office buildings with a height of more than one hundred meters, two super-luxurious hotels, a super-large international mall, international apartments and park. The overall plan closely links the office building, hotel, commercial center and apartment areas. The China Central Place office building is designed with a 45-degree angle facing Chang'an Street, with maximized horizon and natural light. The three office buildings are higher and higher in turn from west to east, outlining the attractive skyline line of the city along the East Chang'an Street. The field of vision on the south is broad without any blocks, and you can enjoy a panoramic view of green land and Tonghui River. It also features a central green area of twelve thousand square meters and a green isolation belt of one hundred miters along Chang'an Street.

The architectural plan is wholly open without columns, providing a clear height of 2.7 meters after elevating the floor to 100 mm. The utilization rate of the whole floor ranges from 71% to 73%.

The fire-fighting system is in line with the international standard, and the automatic fire alarm protecting system is of superfine level. It features perfect smoke control, pressurization and smoke extraction system, and is equipped with reliable emergency lighting system to ensure safety personnel evacuation.

Here the hollowed LOW-E glass curtain wall and advanced air conditioning system create a comfortable office environment. The project also incorporates user-friendly and high-tech elevator system, advanced parking management accounting system, and ladder-styled safety and prevention system with quick support.

华贸中心是北京市2003年60项重大工程之一，是迄今为止唯一横跨东长安街的百万平方米建筑综合体，西临北京CBD核心区，融入并成为CBD的一部分，从机场路国宾道进入长安街，华贸中心是映入眼帘的市区第一道景观。

该项目整体设计由世界著名的KPF建筑师事务所担纲。建筑设计纯净、简洁、明快、凌厉、极富纪念性和时代感。基于"建筑与城市人文结合、建筑与城市环境共生"的理念，总体规划中城市设计的概念将用地东西两面联系起来，使商业自南向北、由密到疏自然地过渡。KPF的设计宗旨是创造一系列的由建筑综合体组成的城市空间，并使这些城市空间的产生与存在与周围环境形成一种必然的联系。

项目由三栋高达百余米的超5A智能写字楼、两座超豪华酒店、超大国际商城、国际公寓和公园组成，总体规划紧密地将写字楼、酒店、商业中心及公寓区联系起来。华贸写字楼呈45度角面对长安街，具有最开阔的视野、最佳的自然采光。三栋写字楼由西向东依次走高，勾画出东长安街沿线充满魅力的城市轮廓线。南面视野开阔无阻，绿地、通惠河尽收眼底，并有12000平方米的中央绿化区以及长安街沿线100米进深的绿化隔离带。

建筑平面采用无柱大开间，在提供100毫米架高地板后净高达2.7米。整层使用率为71%～73%。

消防系统采用国际标准，火灾自动报警系统保护等级为特级。拥有完善的防烟加压和排烟系统及可靠的应急照明系统，以确保人员安全疏散。

中空LOW-E玻璃幕墙、先进的空调系统，创造出舒适的办公环境。建筑中还采用了人性化、高科技的电梯系统、先进的停车库管理计费系统以及阶梯式快速支援、安全防范等系统。

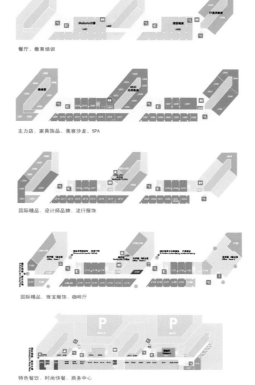

New Shopping Center

GEMDALE PLAZA

北京金地中心

Location: East Dong' an Street, Beijing
Gross Floor Area: 150,000 square meters
Architectural Design: Skidmore, Owings & Merrill LLP (SOM), USA
Developer: Gemdale Group

项目地址：北京东长安街
总建筑面积：150000 平方米
建筑设计：美国 SOM 建筑事务所
开发商：金地集团

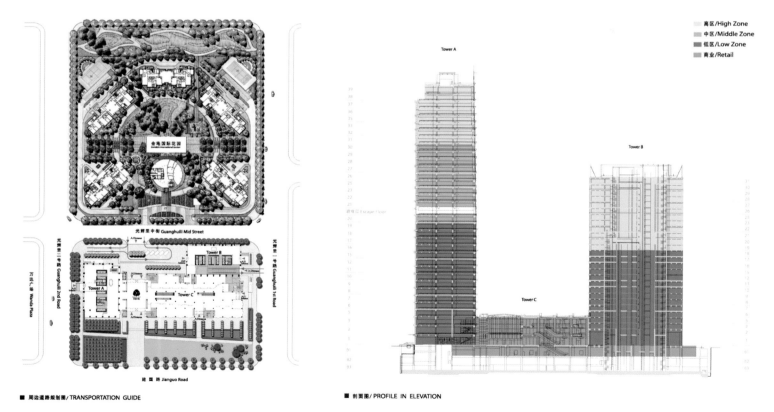

■ 周边道路规划图／TRANSPORTATION GUIDE

■ 剖面图／PROFILE IN ELEVATION

高区／High Zone
中区／Middle Zone
低区／Low Zone
商业／Retail

New Shopping Center | 198-199 | 新商业建筑

Gemdale Plaza is situated at the East Dong'an Street in Beijing, which consists of two buildings that are over a hundred meters. Both the architectural design and interior design are carried by SOM. According to its global designing experience, SOM bears the historical culture of the city, under careful designing; it achieves an organic coordination of both the modern architectural form and the urban historic culture. The one hundred and sixty eight meters' high building group has become the new landmark of Beijing central business area and Chang'an Street.

This project covers a gross floor area of one hundred and fifty thousand square meters, including twenty thousand square meters' business center, the first floor under ground of which connects the No. 1 Subway. Office Building A is one hundred and sixty eight meters high, and Office Building B is one hundred and eight meters high. The light-penetrating glass structured atrium hall takes three floors above the ground which is one hundred and eight meters long and eighteen meters high, connecting the two buildings together as a whole.

On the west side and north side of the office building, by the use of grains and columns, it shows the lattice patterns of traditional Chinese style. The lattice patterns reflected on the dark glass facade further deepen the sense of history. "The lattice patterns", as s kind of architectural language, are frequently appeared as the theme in the design of garden pavement, interior floor and hanging ceiling. On the one hand, such design enhances the concept of the project, and on the other hand, it harmoniously united garden, facade and interior design.

The highly flexible layout can meet different requirements: the areas of the office floors vary from one thousand five hundred square meters to one thousand eight hundred square meters; between the upper and lower floors are the added movable stairs, so that the space can adapt to different functional combinations and layouts. The office building is 2.65 to 2.75 meters high, and takes the best ten to twelve meters depth. It introduces light from four sides, and creates a humane office space.

The office building is equipped with twenty five fast passenger elevators, three fire elevators and four underground garage elevators. The advanced telecommunication infrastructures, along with the information and communication services can support the requirements of knowledge-based enterprises' globalization in such an information age. This project also features twenty-four hour video security monitoring system, state-of-the-art VAV variable air volume system, aerial-floor premises distribution system, satellite communications system, comprehensive mobile communication system, system of Internet wireless in public places, central database system, central autocontrol intelligent system, auto fire alarm system and automatic spraying system. In addition, there are seven hundred and forty five vehicle parking lots and a certain quantity of free parking lots.

Basement 1 地下一层平面图

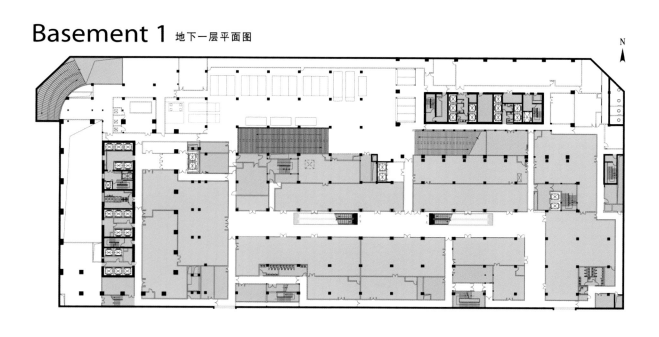

Level 1 一层平面图

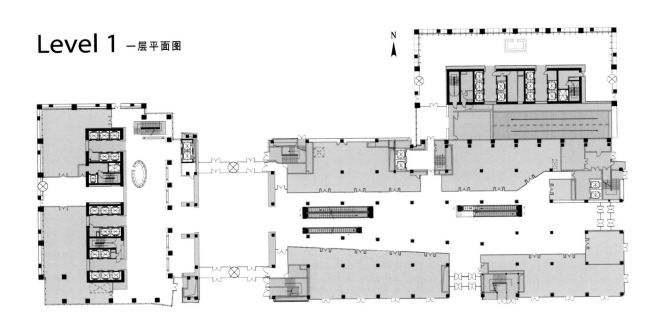

金地中心位于北京东长安街，集两座超百米的高层大厦于一体。建筑设计、室内设计皆由SOM完成。SOM借鉴其全球性的设计经验，传承城市历史文脉，经过细致的设计，使建筑的现代雕塑形式与城市历史文脉有机地协调在一起。高达168米的建筑群，成为了北京中央商务区、长安街的新地标。

该项目总建筑面积150000平方米。建筑面积为20000平方米的商业中心，其地下一层与一号地铁相连通。写字楼A座建筑高度168米，B座建筑高度108米。地上三层的108米长、18米高的全透光玻璃结构中庭大堂在建筑底层将两幢写字楼连为一体。

写字楼的西、北两侧通过梁、柱共同构造出中国古典的窗棂图案。窗棂图案映在深色玻璃幕墙上，加重了建筑的历史感。"窗棂"这一建筑语汇作为母题在园林铺装、室内地面、吊顶等多处重复出现，强化了这一设计理念，并使园林、室外立面、室内效果浑然一体。

高度灵活的布局能迎合多种需求：写字楼楼层面积在1500平方米至1800平方米不等，上下层增设了可拆卸楼梯，适合不同功能的组合和空间布置。写字楼净高2.65米至2.75米，采取10米至12米的最佳进深，四面采光，创造出人性化的办公空间。

写字楼配备25部高速客梯、3部消防电梯、4部地下车库转换电梯。先进的电信基础设施及信息通讯服务支持信息时代知识型企业全球化的商务需求。项目拥有24小时视频保安监控系统、世界领先的VAV变风量全空气空调系统、架空地板综合布线系统、卫星通信系统、全覆盖移动通信系统、公共区域无线上网系统、大厦中央数据系统、中央自控的智能系统、火灾自动报警及自动喷淋系统，另设机动车停车位745个，并提供一定数量的自行停车位。

Level 2 二层平面图

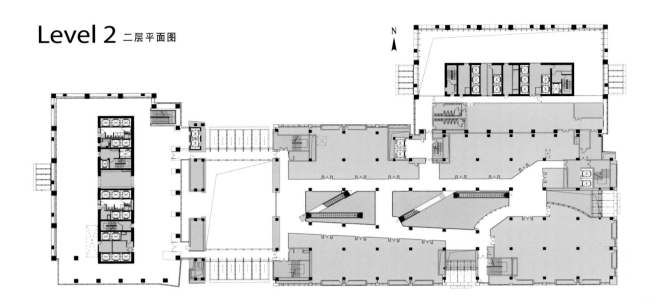

Level 3 三层平面图

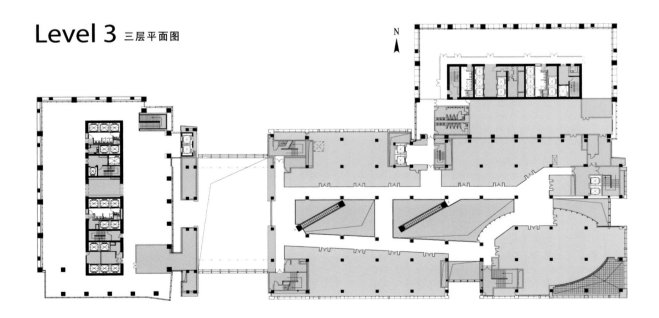

New Shopping Center

VIVA

北京富力广场

Location: West side of Shuangjing Bridge in East Sanhuan Middle Road, Chaoyang District, Beijing
Architectural Design: Aedas Limited (HK)
Developer: R & F Properties

项目地址：北京朝阳区东三环中路双井桥西侧
建筑设计：香港凯达环球建筑设计有限公司
开发商：富力地产

VIVA dominates the south gate of Beijing CBD, closely neighboring the East Third Ring of Beijing. Eight hundred meters to the north, it directly reaches the CBD Trade center; in addition to that, the No. 10 metro line also has a stop here. Having the absolute advantages of location and transportation, VIVA attracts eight hundred thousand people of high consumption ability within the CBD circle and the surrounding 1.5 million stable customers. What is more important is that, with the brand-new concept of international shopping center, it is leading the consumption flow of the whole Beijing, creating the most fashionable and lively Fu Li commercial circle in Beijing.

As one of the most fashionable and charming shopping centers in Beijing, VIVA along with the hotels and office buildings within the complex has created a golden triangle of commerce. It achieves the internal mutual connection and use of commercial resources, and accommodates recreation, entertainment, business, tourism and other nun-shopping elements as much as possible. It fulfills the requirements of one-stop consumption and becomes the focal point of the intersection of business, people, information and money.

This project covers an area of one hundred thousand square meters, containing south area and north area. A 160-square-meter solid envelope shop window, a 490-meter long business frontages in the East Sanhuan road, and a three-storey terrace styled interior interactive entertainment platform, all of these create a shopping paradise featuring business and artificial qualities. This project brings in a brand-new business planning concept, synthetically using the concept of "roof focal platform" + "vertical customers flow direction", thus forming a functional composite mode of attractions at each floor and surprises with each step of the way, and finally this project has achieved a desirable sharing mode of vertical customers flow.

This project sparkplugs the new concept of "six in one SIMPLE LIFE", combining together various business forms like shopping, consulting, club members, parties, recreation and entertainment. It opens up a fresh shopping, recreation and living style, creating a new life concept which integrates eating, drinking, playing, having fun, and shopping into one.

In addition, to better create the commercial atmosphere, this project specially invites famous designers to design the lighting of the facades, striving to make it CBD's most sparkling architecture.

　　北京富力广场雄踞北京 CBD 南大门，紧临东三环，向北 800 米直达 CBD 中心国贸，地铁 10 号线在此设站。北京富力广场具有绝对的地段与交通优势，不仅吸纳了 CBD 的 80 万高消费人群及周边的 150 万稳定客流，还以其全新的国际购物中心理念，引领着全北京的消费流，成就了北京最时尚、最具活力的富力商圈。

　　作为北京最具时尚感与魅力的购物中心之一，北京富力广场与综合体内的酒店、写字楼成就了商业金三角，在内部实现了商业资源互通互用，最大限度地容纳了休闲、娱乐、商务、旅游等非购物元素，满足了一站式的消费需求，成为了商流、人流、信息流和资金流交汇的焦点。

　　该项目整体规模达 100000 平方米，分南、北两区，160 平方米的空中外墙立体橱窗、总长 490 米的东三环商业临街面、三层退台式的室内观赏娱乐互动平台等共同营造了一个商业氛围与艺术气质兼备的购物天堂。项目引入了全新的商业规划理念，综合运用了"空中核心平台"+"垂直客流导向"的概念，以层层吸引、步步惊喜的功能组合模式，达到了理想的垂直客流共享状态。

　　该项目倡导"6 合 1 SIMPLE LIFE"的新主张，将购物、资讯、会员、欢聚、休闲、娱乐等多种商业形态整合起来，开启了一种全新的购物、休闲、生活方式，成就了融合吃、喝、玩、乐、购为一体的新生活理念。

　　另外，为了更好地营造商业氛围，项目还特别邀请了知名设计师进行外立面的灯光设计，力求使其成为 CBD 未来最闪耀的建筑。

BEIJING APM

北京 APM

Location: Wangfujing Street, Dongcheng District, Beijing
Shopping Space: 90,000 square meters
Office Building Area: 40,000 square meters
Developer: Sun Hung Kai Properties

项目地址：北京市东城区王府井大街
商场面积：90000 平方米
写字楼面积：40000 平方米
开发商：新鸿基地产

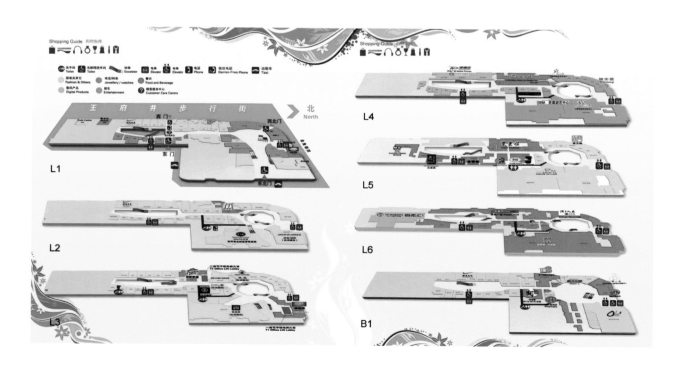

Located in the Wangfujing Street, the golden street of Beijing, the century-old shop New Dong'an Shopping Mall has changed its name to "Beijing APM". This project belongs to Sun Hung Kai Properties Limited Group.

"APM" stands for from "am" to "pm", meaning that Beijing APM will lengthen its business hours, placing more emphasis on the nightclubs, and focuses more on the young and fashionable groups to attract young consumer group from nineteen to thirty-five years old. In order to keep pace with the rapid development of the economy and retailing industry in the mainland and to grab the opportunities of the 2008 Beijing Olympic Games, Beijing APM has invested more than 300 million RMB to go through a large-scaled renovation. The renovated shopping mall adopts large glass screens to invite natural lights to create a more spacious shopping space, and it also places its focus on stressing the images of its business clients.

The brand-new Beijing APM has seven floors, which provides 90,000 square meters of commercial area and 40,000 square meters of A-class office buildings. Beijing APM is bringing in more new super brands and internally-known retailers to further reinforce its leading position. There are one hundred and sixty shops within Beijing APM, among them eighty percent are newly arrived clients. It features various kinds of client combinations, ranging from high-class consumer goods, fashionable dress to entertainment and restaurants, with specially designed areas for jewelry and watches. In the form of independent outlets, it gathers international jewelry and watches of famous brands. The New Dong'an Cinema, the first cineplex in Beijing, now has changed its name to Broadway New Dong'an Cinema, and it has already opened to its customers in Beijing APM.

位于北京金街王府井大街的百年老字号——新东安商场更名为"北京APM"。该项目隶属于新鸿基地产集团。

"APM"的含意为"am"到"pm",意味着北京APM适当延长了营业时间,加重了夜店的比重,以年轻时尚为主线,以吸引更多的19至35岁的年轻消费群体。

为把握内地经济和零售业高速发展及北京2008年奥运会的商机,北京APM投资逾3亿元人民币进行了大型翻新工程。翻新后的商场采用大幅玻璃幕墙引进天然光线,以营造更宽敞的购物空间,并以凸显商户形象的设计为重点。

焕然一新的北京APM共分为7层,可提供90000平方米的商场面积和40000平方米的甲级写字楼。北京APM还引进了许多全新的超级品牌和国际知名零售商,进一步巩固了北京APM的领导地位。内设商铺160家,其中80%为新入驻的商户。商场的商户组合多元化,有高档消费品、时装、娱乐及餐饮等,并特设珠宝、名表区,以独立专门店的形式云集国际珠宝和名表品牌。北京最早的多厅影院新东安影城经过改造后,现已更名为百老汇新东安影城,在北京APM迎接你的到来。

New Shopping Center

INTIME LOTTE

北京乐天银泰百货

Location: No.88 Wangfujing Street, Dongcheng District, Beijing
Gross Floor Area: 83,600 square meters
Business Area: 42,900 square meters
Developer: China Yin Tai Holdings Co., Ltd., Korean Lotte Group

项目地址：北京市东城区王府井大街 88 号
总建筑面积：83600 平方米
营业面积：42900 平方米
开发商：银泰集团、韩国乐天集团

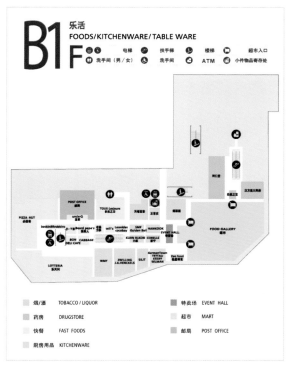

Intime Lotte is built by China Yin Tai Holdings Co., Ltd., a well-known Chinese investment company, and Korean Lotte Group, top five hundred enterprises in the world.

This architecture covers a gross floor area of eight hundred and thirty six thousand square meters, with a business area of forty two thousand and nine hundred square meters. Target at the up-market, this shopping center has a large range of top brand products of many countries, including over five hundred international and domestic brands, such as the international top-class brands GUCCI, ARMANI, Cartier, BVLGARI, and many other foreign famous brands entered Chinese market for the first time. In addition to brand products and fashionable clothing and accessories of different styles, there are also a variety of cuisine and top grade restaurants, which provides customers an easy feast of foods from all over the world while enjoying the fun of shopping. The boutique supermarket on the first underground floor takes the living attitude of being green, happy and creative, offering a high quality and healthy life. The main customer group targeted is the upper class and urban elites who are pursuing a refined life quality and who are with an international fashion taste. This shopping center is offering Chinese consumers a high quality living choice which is more fashionable, more exquisite and more characteristic.

With a good use of advanced Lotte's circulation service system, and run with a brand new managing concept, this shopping center brings in the state-of-the-art business facilities. Intime Lotte also integrates the considerate Korean-style service system with tender Chinese-style services, setting a brand-new service standard for Chinese general merchandise industry. Furthermore, the shopping center specially opens some service programs for distinguished guest, such as the Baby Dayroom, the one-to-one private shopping guide, the MVG (Most Valued Guests) room with professional service staff.

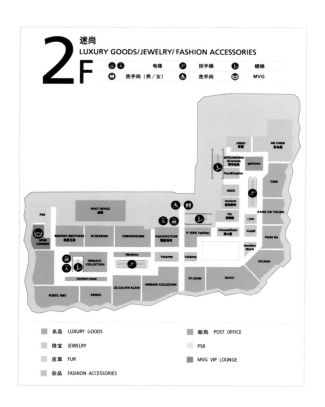

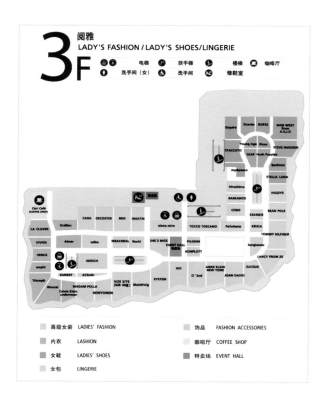

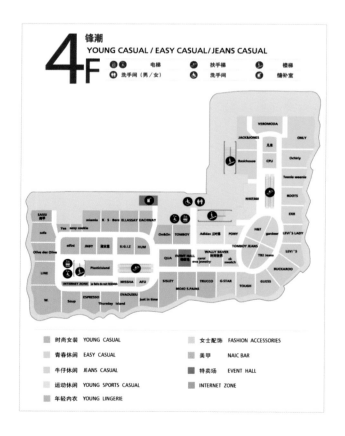
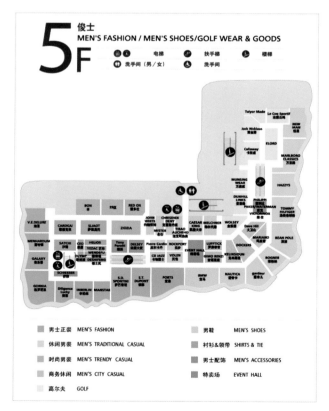

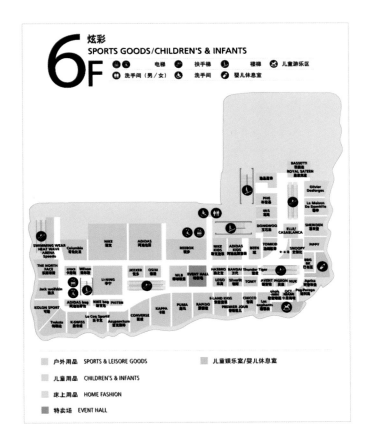

6F 炫彩
SPORTS GOODS / CHILDREN'S & INFANTS

- 户外用品 SPORTS & LEISORE GOODS
- 儿童用品 CHILDREN'S & INFANTS
- 床上用品 HOME FASHION
- 特卖场 EVENT HALL
- 儿童娱乐室/婴儿休息室

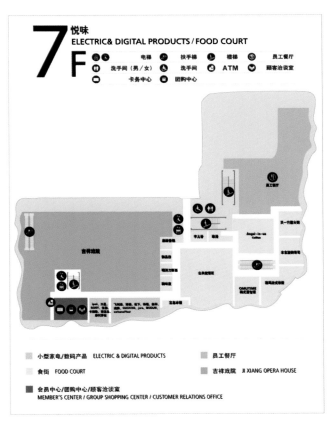

7F 悦味
ELECTRIC & DIGITAL PRODUCTS / FOOD COURT

- 小型家电/数码产品 ELECTRIC & DIGITAL PRODUCTS
- 食街 FOOD COURT
- 会员中心/团购中心/顾客洽谈室 MEMBER'S CENTER / GROUP SHOPPING CENTER / CUSTOMER RELATIONS OFFICE
- 员工餐厅
- 吉祥戏院 JI XIANG OPERA HOUSE

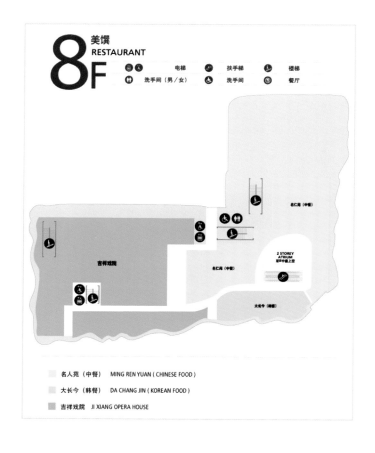

　　乐天银泰百货是由中国著名投资公司银泰集团和世界500强企业韩国乐天集团合资建造的高端百货商场。

　　建筑的总面积达83600平方米，营业面积42900平方米。该商场定位高端，经营多个国家的顶级名品，拥有超过500个国际及国内品牌，其中包括GUCCI、ARMANI、Cartier、BVLGARI等国际一线品牌，以及众多首次入驻中国市场的国外知名品牌。除了不同风格的名品、流行服饰外，更有各色美食及高档餐厅，让顾客在尽享购物乐趣的同时，轻松享受世界各地的美食。地下一层的精品超市撷取绿色、乐活、创意的生活态度，奉献高品质的健康生活。主力客群定位为追求精致的生活品质、具有国际化时尚品位的上流人士及都市新贵。为更多中国消费者带来更时尚、更精致、更具有特色的高品位生活选择。

　　该商场引进国际最先进的营业设施，利用乐天先进的流通服务系统，配合以全新的管理理念进行运营。"乐天银泰百货"还创新地把细致入微的韩式服务体系和脉脉温情的中式服务相结合，为中国百货业的服务树立全新标准。商场还特别开辟了婴儿休息室、一对一私人导购服务、配有专业服务人员的MVG休息室（钛白金会员休息室）等尊贵会员服务项目。

LONGDE PLAZA

北京龙德广场

Location: North Fifth Ring Lishui Bridge, Beijing
Total Site Area: 90,000 square meters
Gross Floor Area: 320,000 square meters
Architectural Design: Woodhead International Pty. Ltd.
Investor: Dalian Yifang Group Co., Ltd., Beijng Grain Group

项目地址：北京市北五环外立水桥
占地面积：90000 平方米
总建筑：320000 平方米
建筑设计：澳大利亚五合国际
投资商：大连一方集团有限公司、北京粮食集团

LONGDE PLAZA is a large-scaled commercial properties project invested and developed by Longde Properties Co., Ltd. which is formed by Beijing Grain Group and Dalian Yifang Group Co., Ltd.

This project enjoys a superb location, and an easy and fast access to transportation facilities. It is located in Ya'ao business circle, one of the six big business circles in Beijing, with over ten public bus routes servicing the area. No.13 light rail and No.5 subway intersect here so one can take the transfer here, too.

This project covers a site area of ninety thousand square meters, with a gross floor area of three hundred and twenty thousand square meters. It is a large-scaled, business-inclusive and high-class shopping mall, which ranges from everyday shop, building material supermarket, general merchandise, household products, catering and recreation, interior furnishings. It is next to the first on the list of shopping malls in Beijing in terms of scale and business area.

LONGDE PLAZA is developed around the business developing mode in the form of orders, oriented with the business combination of household living products, and at the same time, it has both the excellent facades and interior design. Oriented with the forward outlook, combined with urban planning and development, it perfects and enriches the business mode of local shopping centers with a Chinese characteristic, fills up the blankness of the northern Ya'ao area, and completely changes the situation of scattered business distribution, lack of public entertainment facilities of this area; consequently, it changes the shopping habits of the northern Ya'ao area or even the whole northern part of Beijing. It is a rare and precious demonstration project and classic case.

龙德广场是由北京粮食集团与大连一方集团有限公司共同组建的龙德置地有限公司投资开发的大型综合商业地产项目。

该项目地理位置优越、交通便利、可及性高，位于北京市六大商圈之一的亚奥商圈内，有十余条公交线路直达或途经此地，城市轻轨13号线及地铁5号线在此地交叉并可换乘。

项目占地面积90000平方米，总建筑面积320000平方米，规模大、业态种类齐全、高品质，购物中心集生活超市、建材超市、百货、餐饮娱乐、家居及室内饰品等多种业态为一体，在规模和营业面积方面为北京同类第二。

该项目以订单式商业开发模式进行开发，辅以家庭化生活购物中心的商业业态组合进行定位，同时兼有优异的外立面、室内设计。以前瞻性定位，结合城市规划、城市发展，完善及丰富了具有中国特色的区域性购物中心的商业模式，填补了奥北地区的商业空白，彻底改变了奥北地区商业分布零散、公共娱乐设施匮乏的现状，并改变了奥北地区乃至整个京北地区消费者的购物习惯，是一个难得的示范工程与经典案例。

New Shopping Center

YOU-TOWN LIFESTYLE CENTER

北京悠唐生活广场

Location: Chaoyangmen Wai Street, East Second Ring
Total Site Area: 340,000 square meters
Developer: Zhaotai Land

项目地址：东二环朝阳门外大街
占地面积：340000 平方米
开发商：兆泰置地

You-Town City Complex is located in the center of Chaowai commercial circle, beside the East Second Ring. Covering a site area of 340,000 square meters, it is an integrated large architectural complex which integrates residential buildings, business and hotels.

You-Town Lifestyle Center is the commercial sector of the You-Town City Complex. With the rich foreign atmosphere, it attracts millions of high-end consumerist population. Covering 110,000 square meters, this one-stop large comprehensive consumer center has the largest interior central square in Beijing, specialty gourmet confluence, popular entertainment and leisure game world, and the most distinctive sky show wedding auditorium, therefore it became a prosperous sleepless city with fashion spirit in Beijing.

With a business volume of 85,000 square meters, You-Town creates the first square-typed commercial real estate project in Chaowai area. "Square is the living room of a city", and it plays multiple roles, such as reception, communication and exhibition. The commercial square of You-Town is of typical neighborhood layout, in which the overall commercial facilities are organized and marked out within the neighborhood enclosed by the external roads; while at the same time, it is connected to the exterior transportations with the pedestrian areas, so it is easy and independent, forming a good shopping and recreation atmosphere. The natural openness of the square unconsciously leads people to go into the interior space of the building complex.

The facade of the commercial part of You-Town takes western classical style as the principle technique, and the main finishing materials are: dry-hanging granite as the exterior wall, veneer of aluminum alloy and light transparent glass as the curtain wall. With warm-toned stone as the key tone, the skirt building has six six-storey high advertising showcases, so that it avoids the disorder of the building facades caused by the cluttering of advertising after the building complex is used. With an area of 95 square meters, the facade of the Freedom Space gives priority to the veneer of aluminum alloy curtain wall, light transparent glass curtain wall and dark green colored glaze printed glass curtain wall. Floodlighting lamp-houses are set around the top of the roof structure, the walls among the windows of which select the same stone as the skirt building to achieve a unified effect.

悠唐城市综合体位于东二环旁朝外商圈的核心位置，占地340000平方米，是集公寓、商业、酒店于一体的综合性大型建筑群。

悠唐生活广场是悠唐综合体的商业部分，浓郁的涉外氛围引来数以万计的高端消费人群。110000平方米的一站式大型综合消费中心，拥有京城最大的室内中心广场、特色餐饮美食总汇、极具人气的娱乐休闲游戏天地和最具特色的空中show场婚礼殿堂，成为了京城汇聚时尚精神的繁华不夜城。

优唐以85000平方米的商业体量打造出了朝外地区的第一个广场式商业地产项目，"广场是城市的客厅"，肩负着接待、交流、展示等多方面的重要作用。悠唐的商业广场是典型的街坊式布局，整体商业设施都在外围道路围合的街坊内组织规划，采用步行区的方式，同时又和外部交通相联系，便捷而独立，形成了良好的购物休闲氛围。广场天然的开放性，将人们自然地吸引到这个建筑体里面。

悠唐商业部分产品立面以西洋古典式手法为主，主要以干挂花岗石外墙面、单板铝合金幕墙及清色透明玻璃幕墙作为外装修材料；以暖色调石材为主调，裙房共设有六个六层高的广告橱窗，以免建筑使用后由于广告乱设而造成建筑立面的混乱。95平方米的Freedom Space立面外墙主要为单板铝合金幕墙、清色透明玻璃及墨绿色彩釉印刷玻璃幕墙，塔楼顶部周圈设有泛光照明灯箱，塔楼窗间墙选用与裙房相同的石材，以达到整体统一的效果。

New Shopping Center

JIN BAO PLACE

北京金宝汇

Location: No. 88 Jin Bao Street, Dongcheng District, Beijing
Total Site Area: 4,590 square meters
Heigh: 35 meters
Architectural Layers: seven floors above ground, four floors under ground
Parking Floor: the second floor to the fourth floor underground
Gross Floor Area: 40,000 square meters
Ground Floor Area: 26,500 square meters
Underground Construction Area: 13,500 square meters
Shopping Space: 28,000 square meters
Architectural Design: Concept I Design Co., Ltd.
Construction Design: China Architecture Design & Research Group
Developer: Fu Wah International Group, Hong Kong, China

项目地址：北京市东城区金宝街88号
占地面积：4590平方米
建筑高度：35米
层数：地上7层，地下4层
停车层位置：地下2至4层
总建筑面积：40000平方米
地上建筑面积：26500平方米
地下建筑面积：13500平方米
商业面积：28000平方米
建筑设计：国际概念设计有限公司
施工设计单位：中国建筑设计研究院
开发商：香港富华国际集团

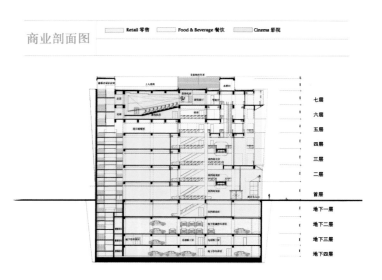

商业剖面图 — Retail 零售 / Food & Beverage 餐饮 / Cinema 影院

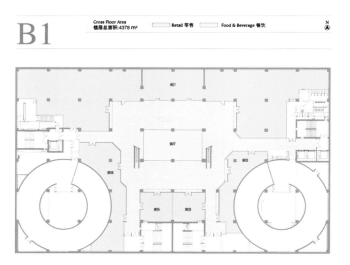

B1　Gross Floor Area 楼层总面积:4378 m²　Retail 零售　Food & Beverage 餐饮

Jin Bao Place Shopping Center sits in the focal place of Beijing Jin Bao Street which enjoys the fame of "World Top Eleven Business Street". As a major part of the "pure up-market consumption circle" of the Jin Bao Street, it is developed and run by Hong Kong Fu Wah International Group. This project covers a gross floor area of over forty thousand square meters, with four floors under the ground and seven floors above the ground.

On the first floor under the ground, there are shops of internationally famous brands. The first floor to the third floor, these floors are for brand clothing, brand watches and accessories. A vast range of top brand flagship shops which focus on dignity and enjoyment gather here. The fourth floor features fashionable living space, providing top-class services. The fifth floor to the seventh floor accommodates top-grade restaurants, international cinemas and recreational parts, where customers can indulge themselves in a light-hearted mood.

Jin Bao Place, with its pure and dynamic artistic appearance, large dry-hanging stone, elegant and fashionable space, echoes with the fine quality of the super five-star hotel and the Hong Kong Jockey Club Institute, Beijing Branch Club, to further lift the image of the entering brands. In contrast to the traditional commercial street, this project gathers world top class brands, integrating up-market shopping, restaurants and recreation into one, including boutique clothing and accessories, brand watches and jewelry, international cinemas, dining and recreation, so that it creates a palace of world-class luxury products for the rich and elites.

This shopping center demonstrates the beauty of briefness in every detail of its interior decoration, and the humane design along with the pure and elegant interior decoration quietly outlines the low-profiled easiness and luxury of the Jin Bao Place to the point. The pure and artistic architectural appearance combined with the fashionable and elegant interior space gives out luxury in a quiet way, and it echoes well with the internationalized aura of the Jin Bao Street.

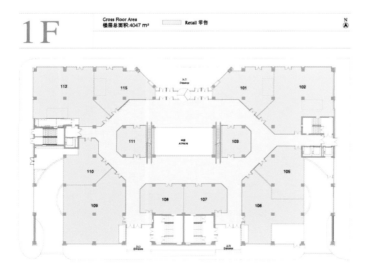

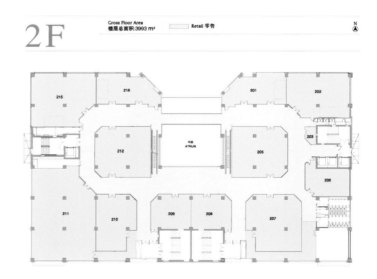

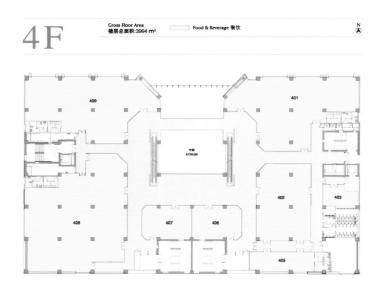

4F Gross Floor Area 楼层总面积:3994 m² — Food & Beverage 餐饮

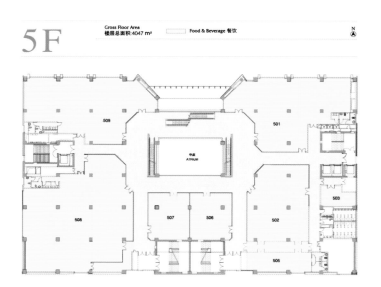

5F Gross Floor Area 楼层总面积:4047 m² — Food & Beverage 餐饮

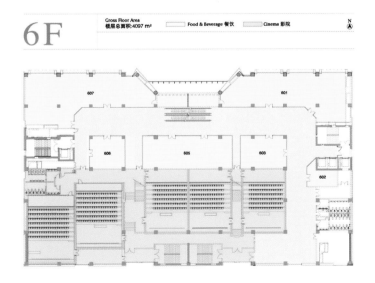

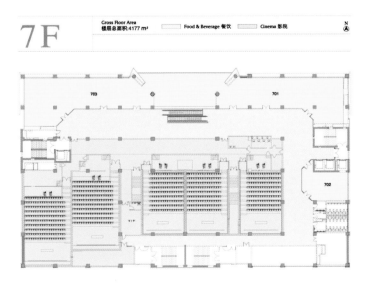

金宝汇购物中心位于"世界第十一商街"北京金宝街的核心位置，作为金宝街"纯高端消费圈"的重要一环，由香港富华国际集团开发与运营。该项目总建筑面积超过40000平方米，地下四层，地上七层。

在建筑的地下一层，诸多国际品牌家居店落户其中。建筑的首层至三层为品牌服装、名表饰品，汇集世界众多一线的品牌旗舰，专注尊荣享受。建筑的四层为时尚生活空间，倡导高端品位服务。建筑的五至七层为高档餐饮、国际影院与休闲娱乐部分，让顾客尽享轻松心境。

金宝汇纯净而富有动感的艺术外观、大幅干挂石材、典雅时尚的空间与两侧的超五星级酒店及香港赛马会北京会所气质相辉映，更加提升了入驻品牌的形象。区别于传统商业区，该项目融汇世界级的一线品牌，集尖端购物、餐饮、休闲娱乐于一体，聚集精品服饰、名表珠宝、国际影城、餐饮休闲，为财富名流阶层筑就世界奢侈品新殿堂。

该购物中心在内部装饰上处处流露着简约之美，人性化的设计以及纯净淡雅的内部装饰毫不张扬却恰到好处地勾勒出金宝汇低调的从容与奢华。纯净而艺术的建筑外观结合时尚典雅的内部空间，将奢华暗自张扬，与金宝街的国际化气质相得益彰。

BEIJING YINTAI CENTER

北京银泰中心

Location: No.2, Jianguomen Wai Street, Beijing
Gross Floor Area: 350,000 square meters
Architectural Design: John Portman & Associates Inc.
Developer: Being Yintai Property Co.,Ltd

项目地址：北京建国门外大街 2 号
总建筑面积：350000 平方米
建筑设计：约翰·波特曼建筑设计事务所
开发商：北京银泰置业有限公司

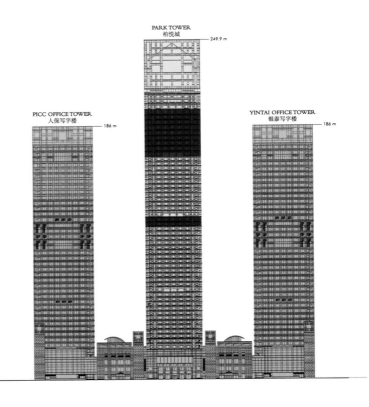

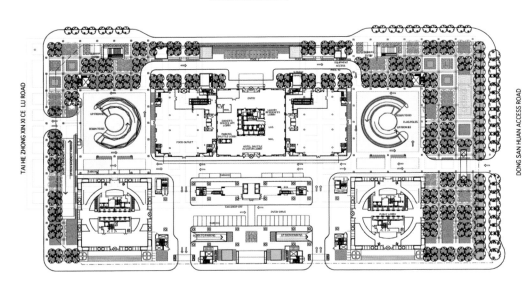

New Shopping Center | 270-271 新商业建筑

Beijing Yintai Center is located in the southwest corner of Guomao Bridge in the East Third Ring, in the core area of the Central Business District, complementing with the International Trade building over a distance, and it is the highest building on Chang'an Street.

The project is designed by world-renowned architect John C. Portman. The elegant and simple design style shows a noble, majestic and generous project, coinciding perfectly with Beijing in the new era. It is the highest building and landmark construction in New Century Chang'an Street. The project covers an area of thirty one thousand three hundred and five square meters, with a total construction area of about three hundred and fifty thousand square meters. It features three main buildings, where the central main building is sixty six floors with sixty three floors on the ground, a total height of 249.9 meters and a pure steel structure. The main buildings include the first Park Hyatt Hotel in China, first-class serviced apartments Park Hyatt Residences and Park Hyatt Residences. The configurations on the east and west sides are symmetrical with 44 floors, and the total height is one hundred and eighty six meters. They are partially steel construction with reinforced concrete structures, forming an intelligent Five-A Grade office building and a perfect podium commercial matching space. The podium area is Park Hyatt life – a new-concept and high-grade fashionable living destination featuring business, leisure, health, food and entertainment. The three square towers are laid out in a triangular structure, showing a tripartite balance of forces. The tower top of the main building is designed with "Chinese Lantern" style, which is a modern interpretation of the traditional Chinese classic patterns.

The project features a three-dimensional landscape design, combining the streets and plants surrounding the buildings, the garden on the podium roof and the interior landscapes. The waterscape design plays an important role in the design of the whole environment, and the designer artistically expresses his understanding of traditional Chinese view on water through a delicate handling technique.

After the opening of Beijing Yintai building, the top of the main building is Beijing Liang Restaurant with 360 degree perspective windows, and you can enjoy the scenery of Beijing Central Business District and the view of Beijing Newtown area.

北京银泰中心位于东三环国贸桥西南角，CBD 中央商务区核心地带，与国贸遥相呼应，为长安街上第一高度。

该项目由世界著名建筑师 John C. Portman 担纲设计，其优雅与简洁的设计风格所体现的高贵、威严、大气，与新时代的北京非常吻合，是新世纪长安街的至高点和地标性建筑。项目占地面积 31305 平方米，总建筑面积约 350000 平方米，拥有三座主体建筑，中央主楼共 66 层，地上 63 层，总高度达 249.9 米，纯钢结构。主体建筑包括中国首家柏悦酒店、极品酒店服务式公寓柏悦居和柏悦府。东西两侧建筑对称配置共 44 层、建筑高度达 186 米，局部钢结构加钢筋混凝土结构，为智能 5A 甲级写字楼和完善的裙房商业配套空间。裙房为柏悦生活——全新概念的高品位商业、休闲、健康、美食及娱乐等时尚生活目的地。三栋方形高塔品字形叠立，呈鼎足之势。主楼塔顶采用"中国灯笼"式的设计，是对中国传统经典图样的现代诠释。

该项目采用立体化的景观设计，使建筑物四周的街道植物、裙房屋顶的花园以及室内景观相结合。水景设计在整体环境设计中起着重要的作用，设计师通过细腻的处理手法将中国传统上对水的认识艺术化地表现出来。

北京银泰大厦开幕后，其主楼顶层为北京亮餐厅，360 度的玻璃窗设计，可观赏到北京 CBD 核心区的风光和北京新城区的风貌。

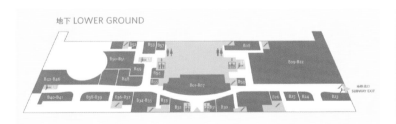

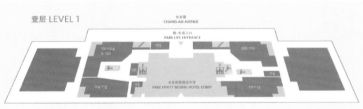

353 PLAZA

上海 353 广场

Location: 353 plaza, No. 353 Nanjing Road, Shanghai
Gross Floor Area: 40,000 square meters
Developer: Gaw Capltal

项目地址：上海市南京东路 353 号 353 广场
总建筑面积：约 40000 平方米
开发商：基汇资本

353 Plaza is the major project of the overall quality improvement of Nanjing East Road. This renovation process maintains the original building facade with the Art Deco style, and converts the interior space of the building into a fresh fashion shopping leisure center, decoding the urban consumption culture in a brand-new perspective. Change your mind, change your urban life, what 353 Plaza on Nanjing East Road has changed is not only just a historic building, but also the fashionable life experiences of the urban new generation, creating a place of their own. 353 Plaza is a multi-functional recreation square which combines clothes and accessories, entertainment, catering and music. The most important component is music. As a result, a music studio is set up in particular on the second floor, which is located in a huge glass capsule. Well-knowned DJs are aperiodically invited to give live performances, and you can enjoy a saxophone performance at a fixed time every day.

In accordance with the special trait of the new generation which is "love music, society, self-expression and individuation", the whole project partitions 353 Plaza into three themed areas: the "fashion code" from the first floor to the third floor becomes the pop information holy land for the new generations; the "trend code" on the forth and fifth floors decodes the trend culture in a new perspective; at the same time, shopping but having fun, the "recreation code" from the sixth floor to the seventh floor makes eating, drinking, playing and enjoying have a "new doctrine". After transformation, 353 Plaza is a new form for urban new generations to redefine shopping and recreation.

353广场是南京东路整体质量提升的重点工程,整修工程保留了原来的Art Deco装饰艺术风格的大楼外观,将建筑内部转形为一个崭新的时尚购物休闲中心,以全新概念解码都市消费文化。改变观念,改变都市生活。南京东路353广场,改变的不只是一幢历史建筑,还改变了都市新生代的时尚生活经验,创造了一个属于他们自己的地方。353广场是集服饰、娱乐、餐饮、音乐为一体的多功能娱乐广场,其中最关键的组成元素是音乐。为此,特别在二楼特设了一个音乐演播室,位于一个硕大的玻璃舱内,不定期地安排知名DJ做现场表演,每天固定时间都有萨克斯表演。

针对新生代"喜爱音乐、社会、自我表现及个性化"的特质,整个项目将353广场规划出3个主题区域:1到3层的"时尚密码"成为了新生代的流行情报圣地;位于4到5层的"潮流密码"则以全新的角度解密潮流文化;同时,购物不忘玩乐,6到7层的"休闲密码"让吃、喝、玩、乐有了"新主义"。蜕变后的353广场,是为都市新生代重新定义购物娱乐的新形态。

New Shopping Center

CITYMALL

北京新中关购物中心

Location: No.19 Zhongguancun Street, Haidian District, Beijing
Architectural Design: P&T Group, BIAD
Developer: Gulfland

项目地址：北京市海淀区中关村大街19号
建筑设计：巴马丹拿国际公司、北京市建筑设计研究院
开发商：海湾地产

CityMall is located in the prime location of Zhongguancun. With over forty bus lines above the ground, and No. 4 and No. 10 metro lines under the ground, the unparalleled accessibility of the CityMall guarantees a massive consumption group to come from the opening of this mall, making this project owning the natural quality of being a bustling commercial circle.

Coving a business area of 47,000 square meters, laying out from two storeys under the ground and five storeys above the ground, CityMall gathers internationally premium brand garments and theme restaurants. It has become a commercial center integrating dining, shopping, recreation and leisure into one. Internationally well-known brand flagship shops have entered this mall one by one, including Warner Brothers International Cinemas, Alexander Hall, and Miss Sixty.

The CityMall introduces the around the clock managing concept. Taking advantage of the large pedestrian flow of the international cinema and the long-hour operation of the subway station, it has realized the aim of extending the peak time of pedestrian flow. By introducing the internationalized innovative concept, it compensates the limitations of the insufficient commercial facilities of Zhongguancun, creating a real 24-hour shopping paradise. In terms of construction design, CityMall embodies the "fashion city with a view" in every corner. To the west side of this project is a small square, a theme square is designed outside the atrium on the second floor, and the upper part of the square is hollow from where the forth floor can be reached. The north side of the square is designed with two sight-seeing elevators in a creative way, where people can enjoy the waterscape on the first floor atrium; as a result, it reveals the dynamic feel of a commercial space all over, leading the crowd to move vertically. The free-streamlined corridors of shops create a fresh sense of shopping and recreation, demonstrating a distinctive visual landscape from different directions and angles.

新中关购物中心商业中心地处中关村核心地带，地上 40 余条公交线路，地下地铁 4 号线、10 号线交汇贯通，无以伦比的可达性使庞大的消费人群如期而至，使项目具备了成为繁华商圈的先天特质。

47000 平方米的商业中心，从地下二层到地上五层，汇聚了国际一线品牌服饰、主题餐饮等，是集吃、购、娱、休闲于一身的商业中心。华纳国际影院、亚历山大会馆、Miss Sixty 等国际知名品牌旗舰店相继鼎力加盟。

北京新中关购物中心引入全天候经营概念，利用国际影院及地铁站长时间营业所带来的大量人流，实现了延长客流集中时间的目的，引进了国际化新颖购物理念，弥补了中关村商业配套不足的缺憾，真正打造了一座 24 小时购物天堂。在建筑设计上，北京新中关购物中心处处体现了"看得见风景的时尚之都"。项目西侧设有小广场，二层中庭处设有主题广场，广场上部中空通达第四层。广场北侧创意性地布置了两部观光电梯，可欣赏首层中庭广场的水景，上上下下中体现了商业空间的灵动感，进而带动人群竖向流动。自由流线的商铺走廊，创造出购物娱乐的新鲜感，从不同的方向和角度，营造出了不同的视觉景观。

参与本书编写的人员：

姚勇、周上高、蒋媛媛、董晗之、杨雪、兰建明、罗明全、杨江、黄严、杨流、杨保、赵辉、胡勇、苏文修、彭运兵、彭运平、孙圣兰、姚太平、陈家、张运梅、胡成兰、胡能、黄红春、何均、杨同力、赵新、谭雄裕、陶海华、姚伟、李国强、张利坤、张迎春、梁富、何有水、李丽华、兰海青、姚健、智超、吴光荣、王桂平、吕柏亮、黄今来、彭丹芬、何安、何平、陈明、罗志强、韩博文、元伟博、康绍辉

参与本书翻译的人员：

邹久娟、王丽红、刘慧敏、范连颖、刘小鹏